FOREWORD

The OECD/NEA Nuclear Science Committee set up in June 1992 a Working Party on Physics of Plutonium Recycling. It deals with the status and trends of physics issues related to plutonium recycling with respect to both the back-end fuel cycle and the optimal utilisation of plutonium. For completeness, issues related to the use of the uranium coming from recycling are also addressed.

The objectives set out are:

- To provide the Member countries with up-to-date information on:

    - The core and fuel cycle issues of multiple recycling of plutonium in partly-loaded LWRs with MOX fuel, and full MOX fuel recycling LWRs cores;

    - The flexibility of fast reactors to produce or burn plutonium within standard fuel cycles (ex.: using MOX fuel), or advanced fuel cycles (ex.: metal, nitride fuels);

    - The core and fuel cycle physics issues related to the use of plutonium fuel without uranium (e.g., plutonium in inert matrices) to enhance plutonium burning capability;

    - The physics issues related to the use of uranium from recycling, and;

    - The core and fuel cycle issues of recycle of plutonium through advanced converter;

- To provide advice to the nuclear community on the developments needed to meet the requirements (data and methods, validation experiments, strategic studies) for implementing the different plutonium recycling approaches.

A co-ordination with complementary studies performed by other task forces/working parties of NSC and by other committees is in place.

The Working Party met three times: once in 1993 to better define the studies to be undertaken and twice in 1994 to discuss results and finalise the reports describing them. About seventy experts from thirteen countries have contributed to this work. The results are summarised in the present volume.

This general volume addresses the issues as seen from the physics point of view and puts into perpective the experience gained. It summarises the results obtained from a series of "benchmark" studies covering physics problems arising from different recycling scenarios. These specific studies, of great interest to the scientific community, are the object of four further volumes.

Finally some further investigations that should answer the questions concerning the physics limitations to multirecycling of plutonium in LWRs are proposed.

This report is published on the responsibility of the Secretary-General of the OECD. The opinions it expresses are those of their authors only and do not represent the position of any Member country or international organisation.

CONTENTS

EXECUTIVE SUMMARY

The future of the nuclear industry will depend crucially on considerations of economics, resource utilisation and environmental impact/waste management. For nuclear utilities who opt for reprocessing/ recycling of spent fuel, one of the main issues pertinent to these three areas is that of recycling plutonium. Moreover, with the recent commitment to reduce the nuclear weapons stockpiles, there has been renewed and generalised interest in the capabilities of both thermal and fast reactors to help make weapons grade plutonium less readily accessible for use in weapons. Additionally there is renewed and widespread interest in the possible role of fast reactors in burning minor actinides. For all these reasons the OECD/NEA Nuclear Science Committee decided to convene an international study group, *the Working Party on Physics of Plutonium Recycling (WPPR)*, to review the physics aspects of plutonium recycle.

The remit of the WPPR was broad, covering most of the plutonium recycle systems, including the mainstream thermal reactors, (Pressurized Water Reactors and Boiling Water Reactors), advanced converter thermal reactors and fast reactors. The Study commissioned benchmark studies in areas it considered to be of particular interest.

- A first set investigated particular issues related to mixed oxide (MOX) usage in PWRs with plutonium both of typical [1] and poor isotopic quality. This last plutonium is expected to become available for recycle early in the next century when multiple recycle of plutonium, as PWR MOX, could be expected to be implemented. The benchmarks were designed to question whether present nuclear data and lattice codes are likely to require further development and validation to be able to satisfactorily calculate the core physics performance with such plutonium.

- A second set of benchmarks examined fast reactor systems to determine the level of agreement as to the rates at which plutonium can be burnt and minor actinides can be fissioned, in particular in order to reduce the source of potential radiotoxicity. Of special interest were studies focused on fast reactor systems which then are optimised for consuming plutonium rather than establishing a breeding cycle, the physics of which has been widely investigated. Such systems may have an important role in managing plutonium stocks until such time that a major programme of self-sufficient fast-breeder reactors is established.

---

[1] e.g., at fuel discharge with burnup of 33 MWd/kg.

The report organisation and main findings are as follows:

Chapter 1 sets the scene for the report and introduces each of the subsequent chapters.

Chapter 2 reviews all the aspects of the physics of plutonium recycle in PWRs and BWRs.

Chapter 3 discusses the findings of the first two benchmarks. The agreement between the various solutions is not completely satisfactory: whatever the benchmark considered, the spread in k-infinity is not less than 1%. Furthermore, this spread in k-infinity could translate in a much larger spread in the plutonium content required to achieve the given reactivity lifetime. This means that there is still a need for improvement in both methods, e.g., self-shielding treatment of Pu-242, and basic data for higher plutonium isotopes and minor actinides. Further experimental validation would also be needed, in particular for integral parameters, e.g., reactivity coefficients, in the case of degraded plutonium isotopic composition. In general, this report underlines that recycle of plutonium in LWRs offers a practical near term option for extracting further energy from LWR spent fuel and reprocessed plutonium. Multirecycling of plutonium in PWRs of current design beyond a second recycling can have intrinsic limitations and the related physics issues have been considered, in particular the plutonium content limitation to avoid positive void effects and the minimisation of minor actinide production during multirecycling. The conclusions reached in this report indicate a good understanding of the physics, although further scenario-type studies (including lattice optimisation) would be needed.

Chapters 4 and *5*, devoted respectively to plutonium recycling and waste radiotoxicity reduction and to the role of fast reactors, indicate that the fast burner reactors / LWR symbiosis offers a potential for significant nuclear waste toxicity reduction by further extraction of energy from the multicycled LWR spent fuel and reprocessed plutonium. However, it has been stressed that for the long term, the best use of plutonium is still in fast breeders. Concerning physics issues, the plutonium-burner fast reactor physics benchmarks display a larger spread in results among participants than has been experienced for more conventional breeder designs. High leakage cores, higher content of minor plutonium isotopes and higher actinide isotopes all need further validation work, including critical experiment performance.

Chapter 6 provides a review of the topic of plutonium fuel without uranium. By using inert carriers for plutonium it is possible to avoid the production of fresh Pu-239 from U-238 captures. This is a topic which has been investigated in the past but is presently of relevance to explore the highest plutonium burnup rates in fast and thermal reactors. New research and development is under way to find potential fuel candidates. The work in this field is closely related to the so-called heterogeneous recycling of minor actinides (i.e. Am) in the form of targets based on inert matrices.

Chapter 7 reviews the possible role of advanced converter reactors as an intermediate step between today's thermal reactors and future fast breeder systems. Advanced converters are thermal reactors in which the conversion ratio is significantly increased above that which conventional thermal reactors can achieve.

Chapter 8 finally provides a brief review of the physics of recycling uranium recovered from spent fuel reprocessing. Although not directly relevant to the subject of this Study, it was felt important to include this for completeness given the close relationship between uranium and plutonium recycle.

Chapter 9 finally gathers the overall conclusions and gives some recommendations for future co-ordinated international work, in particular related to possible limitations in MOX multirecycling in LWRs, nuclear data needs and experimental work.

CONTRIBUTORS TO THIS STUDY

(full affiliations and addresses are provided in Appendix 1)

Chair	M. Salvatores (CEA)
Secretariat	E. Sartori (OECD/NEA)
Editing	K. Hesketh (BNFL)

Chapter Co-ordinators

Introduction
K. Hesketh (BNFL), M. Salvatores (CEA), D. Wade (ANL)

Physics and some Engineering Aspects of Plutonium Recycling in LWRs
H. Küsters (KfK), G. Schlosser (Siemens)

Physics Analysis of Benchmarks for Plutonium Recycling in PWRs
D. Lutz (IKE), G. Schlosser (Siemens), M. Soldevila (CEA), K. Hesketh (BNFL)

Plutonium Recycling and Waste/Radiotoxicity Reduction
M. Salvatores (CEA), D. Wade (ANL)

Plutonium Recycling in Fast Reactors
D. Wade (ANL)

Plutonium Fuel without Uranium
T. Ikegami, M. Koizumi (PNC)

Recycling Plutonium in Advanced Converter Reactors
R. Jones (AECL)

The Use of Recycled Uranium
R. Jones (AECL)

Conclusions and Recommendations
M. Salvatores (CEA), E. Sartori (OECD/NEA)

Benchmark Specification

Plutonium Recycling in PWRs
G. Schlosser (Siemens), J. Vergnes (EdF), H. W. Wiese (KfK)

Void Reactivity Effect in PWRs
Th. Maldague (Belgonucléaire), G. Minsart and P. D'Hondt (SCK-CEN)

Plutonium Burner Fast Reactor
J.C. Garnier (CEA), T. Ikegami (PNC)

Metal Fuelled Fast Reactor
D. Wade (ANL)

Benchmark Results Analysis and Report Editors

Vol. 2. *Plutonium Recycling in PWRs*
    D. Lutz, A. and W. Bernnat (IKE), K. Hesketh (BNFL), E. Sartori (OECD/NEA)

Vol. 3. *Void Reactivity Effect in PWRs*
    D. Lutz, A. and W. Bernnat (IKE), K. Hesketh (BNFL), E. Sartori (OECD/NEA)

Vol. 4. *Fast Plutonium-Burner Reactors: Beginning of Life*
    G. Palmiotti (ANL)

Vol. 5. *Plutonium Recycling in Fast Reactors*
    R. N. Hill and K. Grimm (ANL)

Text Processing and Report Outlay

P. Jewkes (OECD/NEA)

Benchmark Participants

Organisation	Names	1	2	3	4
AEA Technology	Smith				*
ANL	Blomquist	*	*		
	Palmiotti			*	
	Grimm, Hill				*
Belgonucléaire	Maldague	*	*		
BNFL	Mangham	*			
CEA	Puill, Marimbeau	*			
	Aigle, Cathalau, Maghnouj, Soldevila		*		
	Rahlfs, Rimpault		*		
	Garnier, Varaine			*	
	Rimpault, DaSilva				*
SCK-CEN	Minsart		*		
ECN	Aaldijk, Freudenreich, Hogenbirk		*		
	Wichers		*		
EDF	Barbrault, Vergnes	*			
ENEA	Landeyro		*		
Framatome	Kolmayer	*			
Hitachi	Ishii, Maruyama	*	*		
IKE	Bernnat, Käfer, Lutz, Mattes		*		
	Lutz	*			
IPPE	Tsibulia		*	*	*
JAERI	Akie, Mori, Okumura, Takano		*		
	Akie, Kaneko, Takano	*			
PNC	Ikegami, Ohki, Yamamoto			*	*
PSI	Holzgrewe, Paratte	*			
	Pelloni			*	
Siemens	Hetzelt, Schlosser	*	*		
Studsvik	Ekberg	*			
Toden	Saji	*			
Toshiba	Uenohara		*		
	Kawashima, Yamaoka			*	

Chapter 1

INTRODUCTION

The future shape of the nuclear industry world-wide will be strongly influenced by three major considerations – economics, optimal use of finite resources, and environmental impact/waste management:

- *Economics*

    There is generally increasing pressure on nuclear power to be genuinely economically competitive with other large scale energy sources, such as coal, oil and gas. Developments such as improved coal burning systems and combined gas cycle power plants are changing the economics of the fossil sector and increasing the pressure on nuclear utilities to reduce capital and operating costs while not compromising safety.

- *Resource utilisation*

    Another factor which is becoming increasingly important is that of making the best use of the world's finite energy resources. The important considerations here are increasing power plant efficiency, energy conservation, improved energy efficiency and improved fuel utilisation.

- *Environmental impact/waste management*

    There is increasing public awareness of the importance of proper accounting of the environmental impact of energy production and use. The emphasis is on global analyses which account for all of the environmental effects of energy production wherever they occur and not just on the impact on the area local to a power plant. This is an area where it can be very difficult to quantify all of the environmental costs and benefits, making comparisons between different strategies very difficult. A crucial element of minimising environmental impact is the responsible management and minimisation of all waste arisings and there is increasing understanding of the need for planners to take proper account of the full costs to society of waste management/disposal in intercomparing the costs of energy production among various alternatives. No matter what the energy source, past practice has often been to neglect the full environmental and health costs of waste disposal, an approach whose shortcomings are now recognised. For example, disposal of CO_2 to the atmosphere carries no costs in most analyses of fossil plants, while decommissioning costs are often ignored for plants other than nuclear.

Moreover, the issues of **safety and non-proliferation**, while not quantifiable like economics, resource utilisation and environmental impact, must be accounted for because of the necessity for the widespread public acceptance of any options selected for the future nuclear industry. It hardly needs saying that the safety of all activities associated with nuclear power is of paramount importance. Although some of the physics and engineering aspects referred to in this report impinge on questions of safety, detailed discussion falls outside the scope of this Study. As with safety issues, the question of non-proliferation also falls outside the scope of this Study. All of the plutonium recycle activities described in this report will need to conform to the highest standards as regards non-proliferation and verification.

The management of spent fuel from nuclear reactors and a full understanding of the underlying scientific issues are crucial to all of the above areas. There are a number of options for spent fuel management, ranging from interim storage followed by direct disposal of the intact fuel assemblies to chemical reprocessing of spent fuel assemblies and recycling of some or all of the fissile materials recovered in either thermal or fast reactors or via a combination of both. There is at present no clearly accepted best option and there is a tendency for each country and indeed each nuclear utility within a country to have its own viewpoint and its own preferred strategy. It is therefore essential to clarify as far as possible the scientific issues related to the different options.

Regarding the economics of nuclear power, the choice of spent fuel management strategy has a modest but nevertheless significant impact on overall generation costs. Economics of plutonium and uranium recycling have been well studied and an account can be found in recent OECD publications (Plutonium Fuel: An Assessment, OECD/NEA, 1989 and Economics of the Nuclear Fuel Cycle, OECD/NEA, 1994). Only a few facts are reminded here that help putting into perspective the physics issues discussed in the present report. Spent fuel management normally constitutes about 15% of the fuel cycle costs of a thermal reactor such as a Pressurised Water Reactor (PWR) or a Boiling Water Reactor (BWR), although it can be substantially larger in countries whose projected costs of waste management facilities are particularly high. Although spent fuel costs are very large in absolute terms, their impact on overall generation costs are very modest, bearing in mind that fuel cycle cost component of modern PWRs and BWRs typically only represents about 20% of the overall cost (the other 80% being made up of capital charges for the nuclear power plant, which is the dominant cost and operating and decommissioning provisions).

The spent fuel management strategy which a nuclear utility opts for in the context of a national policy has a direct influence on resource utilisation. In fact it is well known that a once-through thermal reactor fuel cycle with ultimate direct disposal of spent fuel does use only part of the original uranium ore (approximately 1%). In order to conserve uranium reserves (if this objective is of relevance), recycle fissile materials from spent fuel may be envisaged so that increased amounts of energy from the uranium ore are extracted. This may extend to re-use of the residual U-235 and fissile plutonium in spent fuel in thermal reactors to recycling of these products in fast reactors. Studies show that the impact of the various options available on resource utilisation ranges widely, from a modest figure of perhaps 30% increase in energy output per kg of uranium ore to the case of one-time recycle in thermal reactors to a factor of up to 100 times in the case of extended recycle in fast reactors.

Finally, there is the issue of the environmental impact of spent fuel management. A nuclear power plant operating normally retains all but a minuscule fraction of the fission products and transuranics in the spent fuel. The spent fuel management option chosen by the utility determines the eventual fate of these radioactive by-products. As for plutonium, its presence in spent fuel or radioactive waste intended for geologic disposal has a significant impact on the long-term potential radiotoxicity source.

Underlying all the issues touched on so far are the technical questions and the physics issues relating to the recycling of plutonium. In line with the discussion above, attitudes to plutonium management vary from country to country and from utility to utility. Some see plutonium as a constituent of spent fuel which is best committed to direct disposal along with the intact fuel assemblies.

Others with past or future reprocessing commitments vary in their attitude to the plutonium. Some regard plutonium as a liability and others as a valuable energy resource. The new issue of the use or disposal of ex-weapons plutonium is a further complicating factor with many wide ranging implications, although it forms only a small fraction of current world-wide reserves.

This Study was initiated with the intention of comprehensively reviewing the current status of the physics of plutonium recycle, as a contribution to a wider understanding of the broader issues discussed above. It complements another recent study carried out in the frame of the Nuclear Development Committee, and being published by the OECD/NEA on the available technologies for plutonium management.

The physics of plutonium recycle is a broad and complicated technical area some aspects of which are well understood and well proven, while others may be uncertain to a greater or lesser extent or even very speculative. The purpose of this Study is to review the physics of plutonium recycle as it stands today and to identify what tasks remain to be done to support future plutonium recycle strategies. Although it is not the purpose of the Study to fully examine all aspects of plutonium recycle such as waste management, proliferation and risk, it is not practical nor is it appropriate to entirely divorce the physics issues from them. Consequently this Study will specifically address amounts, compositions and toxicities of plutonium and transuranic flows and inventories as they are impacted by various choices for managing the back-end of the fuel cycle. Also addressed explicitly are the impacts of reactor safety and safety of alternative systems.

The various chapters of this report consider the following broad areas:

- *Thermal reactor plutonium recycle*

  With the large scale deployment of fast reactors a distant prospect in most countries, many nuclear utilities are planning, or have already implemented plutonium recycling schemes in thermal reactors, principally PWRs (though there is limited experience and plans for future plutonium recycle in BWRs). Plutonium is normally recycled as Mixed Oxide (MOX) fuel in which it is mixed with uranium dioxide (UO_2) in the form of plutonium dioxide (PuO_2).

  Regarding the use of MOX derived from current reprocessing plants in PWRs at least, the physics aspects are essentially fully understood and give rise to no operational, licensing or safety concerns in those commercial power plants where it is already in use. Uncertainties are always present, of course, but their importance increases in significance when considering future MOX fuel cycles with higher discharge burnups and consequently higher initial plutonium contents. This is compounded when the plutonium itself may be recovered from spent MOX fuel assemblies which have themselves been reprocessed and the isotopic quality of the plutonium (meaning the ratio of thermally fissile isotopes Pu-239 and Pu-241 to total plutonium) is degraded. This is a situation which is likely to arise in some countries with mature programmes of thermal reactor recycle starting sometime beyond the turn of the century. Since this is the area where the greatest unknowns and uncertainties lie, the issue of multiple recycle of plutonium in thermal reactors is a central theme of this Study and forms the basis of the thermal MOX benchmark exercises described in Chapter 3.

  In the context of multiple recycle of thermal MOX, the question most often posed is "how many times can thermal MOX be recycled". This question which is discussed in Chapter 2 is of particular relevance. In fact, each time thermal MOX is recycled, the isotopic quality of the plutonium degrades and there will come a point where its further recycle in thermal reactors is not practicable. Even if all the physics issues associated with multiple recycle were understood perfectly, this question could not be answered simply, because it depends on the particular scenario, especially on the ratio in which UO_2 and MOX assemblies are blended in the reprocessing operations. In current thermal reprocessing plants it is not desirable to process campaigns made up entirely of MOX assemblies and the intentions are to admix MOX assemblies with UO_2. This delays the degradation of the plutonium and allows an increased number of recycles to be achieved. Other factors affecting the number of thermal recycles are the reactor type, the fuel design and the discharge burnup, all of which would contrive to complicate the answer.

A simpler approach to the question is, however, possible. It involves asking not how many times MOX can be recycled, but what is the maximum concentration of plutonium in thermal MOX that is practicable. Although the exact answer is not yet known, this Study has concluded tentatively that the upper limit lies in the range 10 to 12 w/o *total* plutonium, assuming thermal MOX designs similar to those already in use today. For blending ratios foreseen in today's reprocessing plants (typically three or four UO_2 assemblies blended with every MOX assembly), this would permit *at least* two recycles of plutonium as thermal MOX. With a higher blending ratio or with a modified lattice geometry, an increased number of recycles could be possible, but the precise number is not yet determined. Moreover, actinide buildup during multiple recycling, whatever the moderator-to-fuel ratio, will increase and can become a potential criterion for multirecycling feasibility. This point is discussed in Chapter 4.

- *Thermal benchmark exercises*

  Benchmarks relevant to the recycling of plutonium in PWRs were specified as part of the Study. One of these was a wide comparison of lattice code predictions for a simple LWR pin cell with plutonium of good isotopic quality representative of that coming from today's commercial reprocessing plants. For this case there exists a large amount of crucial validation evidence in the form of experimental measurements from zero power critical facilities and operating experience in power reactors. It was included so as to act as a baseline against which to compare a similar exercise in which very poor quality plutonium similar to that which might arise in future with multiple recycling in PWRs (i.e. when PWR MOX assemblies are themselves reprocessed and recycled) and for which there is no validation at present. Finally, a void effect benchmark was specified to determine whether the various physics codes available world-wide would agree as to the sign of the moderator void effect in a hypothetical and computationally challenging voiding pattern in a supercell containing a MOX assembly also containing poor quality plutonium. The analysis of these benchmarks is detailed in Chapter 3.

- *Plutonium recycling in fast reactors*

  Chapter 5 considers the physics of plutonium recycle in both oxide fuelled fast reactors (with conventional PUREX-based recycle) and metal alloy fuelled fast reactor systems (with PYRØ-based recycle). Two topics of particular interest today are the potential of fast reactors for consuming plutonium and fissioning minor actinides because *burner* reactors have not received the same level of focus in the past as *breeder* reactors. These two topics underlie the fast reactor benchmark exercises which were devised as part of this Study. A major driving force for the Study was the lack of international validation in this area.

- *Fast reactor benchmark exercises*

  Benchmark exercises were defined for both oxide fuelled fast reactors and metal alloy fuelled fast reactors. The purpose was to establish the level of agreement between different fast reactor code systems regarding the rate of destruction of actinides and plutonium in particular. A novel aspect of the benchmarks is the departure from the usual assumption of a fast reactor breeder – cases with non-breeding fast reactor systems (with conversion ratios in the range 0.5 to 1.0) were considered. At least as far as a breeding ratio of 0.5 is concerned, existing fast reactor designs could easily be re-designed to become non-breeders simply by omitting the radial breeder region.

In addition to the two main areas above, this report also looks at two other aspects of plutonium recycle and also the topic of uranium recycle:

- *Waste and radiotoxicity reduction*

  Recycle of plutonium has an important impact on the waste arising and on its radiotoxicity. In the long term, the radiotoxicity is dominated by the arisings of transuranics which can differ significantly depending on the plutonium recycle strategy. In particular, fast reactors can be used to achieve considerable reductions in the transuranic arisings per unit of electricity production. Chapter 4 considers the physics basis underlying this.

- *Plutonium fuel without uranium*

  There has recently been increasing interest in using specially designed thermal reactors to optimise the management of civil plutonium stocks. The objective is to remove the plutonium from store and place it in-reactor where it is inherently more safeguardable. There is also a great deal of interest in the possibility of using ex-weapons plutonium for reactor fuel as a means to decrease the world stockpiles of plutonium pits. The concept of plutonium fuel with a carrier other than uranium has been extensively studied in the past and there is renewed interest since it offers the potential for incinerating plutonium by fission without at the same time generating any new Pu-239 from U-238 captures. Chapter 6 covers both thermal and fast reactors with plutonium fuel without uranium.

- *Recycling of plutonium in advanced converter reactors*

  Chapter 7 provides a brief review of advanced converter reactors and their relevance to plutonium recycle. These are thermal reactors which are optimised to achieve high conversion ratios and, in the case of heavy water moderated variants, low fissile inventories. The low critical mass can allow the fissile content of discharged fuel to be burnt down to the point where recovery of the residual fissile material is not worthwhile. Advanced converter reactors therefore represent an intermediate step between today's thermal reactors and fast breeders, either as the final irradiators before disposal, or as an "active store" for plutonium in the interim period.

- *The use of recycled uranium*

  Chapter 8 provides a brief review of the physics of recycled uranium. Although this is not directly relevant to the topic of the Study, this is included for completeness because not only plutonium but uranium as well is a product of recycle.

Chapter 2

PHYSICS AND SOME ENGINEERING ASPECTS
OF PLUTONIUM RECYCLING IN LIGHT WATER REACTORS

This chapter deals with the neutron physics aspects of plutonium (Pu) recycling in PWRs and BWRs. In addition, such engineering aspects as mechanical and chemical properties of MOX are also discussed. The results of the physics investigations are compared with measurements from critical experiments and measurements from LWR power reactors. The issue of multiple recycling of Pu in LWRs will need further investigation in due course.

2.1 Introduction

Considerable amounts of Pu are generated in uranium fuelled light water reactors (LWRs). About 220 kg of fissile Pu are available per 1 GWy from discharged fuel. Pu recycling in LWRs is beneficial with respect to resource, environmental and safeguards considerations. This chapter presents a review mainly of the reactor physics aspects of Pu recycling in LWRs. In particular, it identifies those aspects of reactor physics which are already well proven and those which have yet to be fully established, such as multiple recycling. Economic questions are outside the scope of this review, but the experience of recycling of Pu in LWRs is one of the main items.

We will deal specifically with Pu generation and consumption in LWRs and high burnup fuel utilisation concepts. The investigation of neutron physics aspects will be the central part of this Chapter. MOX design problems and verification of Pu recycling in PWRs and BWRs via benchmarks, critical experiments and with LWR reactor cores containing MOX will be discussed in detail. A consideration of material properties of MOX fuel as regards fabrication and reprocessing will also be given.

This overview will summarise the present status of various Pu recycle concepts, their technical feasibility and the practical limits in reducing Pu quantities by burning in LWRs.

2.2 Plutonium utilisation concepts in LWRs

2.2.1 Plutonium generation in LWRs

As there is no significant difference in the neutron spectra of different LWRs, the Pu production at present day burnup values (up to about 45 MWd/kg) is about 220 kg fissile Pu (Pu-239 + Pu-241) per GWy. The final burnup of unloaded UO_2 fuel assemblies governs the isotopic composition of the Pu as is shown in Figure 2.1 for a fuel assembly with an initial enrichment of 4 w/o U-235 in a modern PWR. Only the Pu-239 concentration decreases with irradiation from its highest value, because in-situ production of fresh Pu-239 is insufficient to offset the removal rate. In contrast, the concentrations of the other Pu isotopes increase with burnup, reflecting the high rate of neutron captures. Of principal interest for Pu recycling are:

- The content of thermal fissile Pu isotopes Pu-239 and Pu-241,

- The disadvantage for nuclear design of increasing amounts of thermally non-fissile Pu-240 and Pu-242,

- The increase of Pu-238 and Pu-241.
  These are the main sources of radiation (neutrons from Pu-238, gammas from decay of Pu-241 via Am-241) and decay heat (from Pu-238), which are important considerations for the fabrication and handling processes.

2.2.2 Plutonium utilisation strategies

Since any design of MOX fuel assemblies and MOX containing cores has to obey the same safety requirements as UO_2 cores, Pu-bearing fuel rods and fuel assemblies have to meet the same thermal and mechanical limits as specified for UO_2 fuel.

Specific additional costs associated with the fabrication of MOX fuel, which are thought to be significantly reduced at increased throughputs, give an incentive to select the Pu concentrations as high as possible consistent with nuclear and thermal limitations.

Related to the MOX content of the cores the following cases are of interest [1]:

- *Self generated recycling* (SGR) assumes that only a Pu quantity equivalent to that previously generated in the same power station will be recycled, allowing for a delay of some years following discharge for pond cooling, reprocessing and fabrication;

- *Open market recycling* (OMR) allows an earlier start to recycling through the availability of Pu from other reactors and also allows higher MOX fractions in the core than is possible with SGR;

- A thermal Pu burner with 100% MOX fuel (i.e. without UO_2 fuel assemblies) is the limiting case of OMR. As there are no UO_2/MOX transition effects to be concerned about, this allows the MOX fuel assembly and core designs to be optimised and simplified and still satisfy the safety requirements, though there may be requirements for upgraded control rod and boron systems.

2.2.3 MOX fuel assembly concepts for PWRs and BWRs

To make the use of Pu in LWRs as competitive as possible there is not only the need to concentrate the Pu in the fewest number of fuel rods but also in the minimum number of fuel assemblies to minimise the additional costs for fabrication and transportation.

Interface effects between UO_2 and MOX have very important influences for the design of MOX assemblies for both PWRs and BWRs. For the latter there are also the water gaps between assemblies and inner water areas inside assemblies, both of which have important additional influences on the MOX assembly design.

Two assembly configurations have been investigated and tested in the past [2]:

- *The "plutonium island" assembly.* This is an assembly with the MOX fuel rods located in a central zone and enriched uranium rods at the periphery. This concept appears more suited for BWRs or any reactor having large water gaps between assemblies;

- *The "all plutonium" assembly.* This is an assembly comprising MOX fuel rods only. It is more appropriate for PWRs, where the effect of guide-thimbles on flux and power peaking are corrected by an adequate choice of enrichment zoning. New proposals for BWRs use this configuration type, too.

In the centre part of a MOX rod area the plutonium content is selected high enough to satisfy the required reactivity and burnup. There is a tendency to have rather large areas of this kind. This favours fuel assemblies with MOX rods only, where the fissile plutonium content must be lowered only in the outer rods to avoid the power peaking induced by the thermal flux of the surrounding uranium fuel and moderating water areas.

Up to now, a significant fraction of MOX fuel assemblies have used natural uranium as the carrier material. Depleted uranium produced downstream of uranium enrichment plants as a waste product (tailings) with a U-235 concentration of 0.2 to 0.3 w/o is now considered a more attractive carrier material and is used in more than 60% of all MOX assemblies fabricated. Owing to their low fissile uranium content, uranium tailings offer the opportunity of maximising the plutonium content.

There is a general tendency toward increased burnup levels. This is especially advantageous for the economics of MOX fuel assemblies and leads again to higher fissile plutonium concentrations.

Figure 2.2 and Figure 2.3 show examples of the design of MOX fuel assemblies in use in German [3,4] and French [5] PWRs. Figure 2.4 shows the MOX assemblies fabricated for insertion in Gundremmingen B/C. Finally Figure 2.5 shows a MOX assembly with improved water structure for future use in BWRs [3,4]. Also undergoing the licensing process are MOX fuel assemblies (FA) designed by SVEA which give improved inner moderation.

Hexagonal MOX fuel assemblies proposed for WWER-1000 reactors also incorporate three regions of different Pu content. To achieve a burnup of about 40 MWd/kg, Pu contents of 1.6/4.0/6.3 w/o are required in regions of 66/48/198 fuel pins respectively, using U-tails as carrier [6].

2.2.4 Improved fuel utilisation, multirecycling strategies

The present trend is to increase the fissile content of UO_2 fuel assemblies to reach higher burnups. This trend is of interest for the MOX fuel assemblies, too, especially since the fabrication cost is essentially independent of burnup . On the other hand, as a consequence of the in situ burning of Pu, the resource utilisation only improves slowly with burnup.

The concept of a closed fuel cycle with reprocessing and recycling implies also the reprocessing of MOX fuel assemblies [3]. A separate reprocessing campaign for MOX would result in the extraction of poor quality plutonium (low fissile content, higher Pu-238 content) from the fuel cycle. Present intentions, however, are to mix all available plutonium, thereby avoiding such a rapid degradation of plutonium quality. At the present time, the optimal strategies for multirecycling have yet to be clearly defined (reactivity coefficients, limitations on minor actinide production, etc.). Additional studies of the nuclear conditions are needed in this context.

The resulting changes in the plutonium inventory of a PWR fuel cycle without and with recycling up to the third generation as function of time are shown in Figure 2.6 [7.

2.3 Neutron physics

2.3.1 General aspects [2]

Within the neutron spectrum typical of LWRs, Pu-239 and Pu-241 are the only fissile isotopes, although the other isotopes are fissile for high energy neutrons. Due to neutron-gamma reactions in competition with the fission reactions, the various plutonium isotopes are transmuted into plutonium isotopes of higher atomic mass. This coupled chain, containing two fissile isotopes separated by a fertile isotope, results in a variation of reactivity with burnup which is much flatter for MOX than for UO_2 fuel. As a result, the reactivity of a core containing MOX fuel assemblies decreases less rapidly with burnup than that of a core containing initially only UO_2 fuel, providing better stretch-out capabilities.

The variation of the cross-sections of plutonium isotopes with energy is more complex than for the uranium isotopes. The absorption cross-sections of the main isotopes (Pu-239, Pu-241) are about twice as important as for U-235 in the thermal energy spectrum, resulting in a relatively smaller reactivity value for the control rod worth or for the boric acid in a UO_2-PuO_2 lattice. In addition, the absorption cross-sections of the plutonium isotopes are characterised by absorption resonances more numerous and much more important in the epithermal energy range (0.3 to 1.5 eV) than those of the uranium isotopes. Moderator temperature and fuel temperature (Doppler) coefficients are therefore more negative for MOX fuel than for UO_2 fuel.

The overlapping of all the plutonium isotopes and all their resonances makes the analysis of UO_2-PuO_2 lattices a challenge.

The rather small differences between the isotopic composition of fission products from plutonium and uranium isotopes have multiple consequences, but generally result in lower overall fission product activity.

The fission yield of iodine and its precursors is slightly higher from plutonium fissions than from uranium fissions. Besides the environmental impact, this feature might be expected to lead to a higher propensity for stress corrosion cracking and therefore pellet clad interaction failure. This is balanced by the better creep properties of MOX fuel observed experimentally.

The tritium fission yield is also increased. Under normal power plant operating conditions, the environmental impact is small for PWRs and negligible for BWRs. It may even be nil or beneficial if burnable poisons have to be used as a result of recycling plutonium.

Xe-135 is a strongly absorbing fission product, affecting the operation of a LWR in such areas as the reactivity margin (to allow for reactor start up within the hours following a shutdown), xenon peaks, and power distribution oscillations. It has approximately the same yield from uranium and from plutonium fissions. However, due to the higher absorption cross-sections of MOX fuel at thermal energies, the reactivity effect and power distribution oscillations are reduced. A plutonium recycle core is thus more stable with regard to xenon oscillations than current LWR cores.

The fraction of delayed neutrons [8,9] produced by Pu-239 is only about one-third of the fraction produced by uranium isotopes. For Pu-241 the fraction of delayed neutrons is about the same as for U-235 (see Table 2.1). Consequently the contrast between MOX cores with respect to UO_2 cores is reduced for Pu of highly burnt fuels and is of no significance for control and safety behaviour.

The activity inventory of equivalent UO_2 and MOX fuel assemblies at the same power and burnup is determined by the fission products. As stated above the differences in fission product yields are small. For short times after shutdown and in accident conditions, the MOX fuel assemblies (FA) show a somewhat reduced activity for fission products and actinides. The decay for MOX is less steep, however, and so the advantage in favour of MOX has vanished by the time of unloading from the reactor core. For very long final storage times of burnt FAs or their reprocessing wastes, the contribution from the decay chains of actinides (such as Am-243, Cm-244, and Np-237) becomes dominant compared to fission products.

Due to its long half-life, Np-237 dominates the toxicity of highly active waste after about 10 000 years. Reactor designs with a reduced production of this isotope may be of interest. Some reduction could be reached by increasing the moderation ratio of the MOX fuel lattice. As significant quantities of Np-237 result from the decay of Pu-241 via Am-241, short out of core times should be sought as far as possible, too.

In the closed fuel cycle the activity level in the long term is less than that of the open fuel cycle because of the lower quantities of Pu following MOX recycle. Although minor actinides enhance the activity initially, their half lives are generally fairly short and in the long term the plutonium dominates.

A comparable behaviour can be found for the decay heat. The storage of burnt MOX-FAs in the power station leads under very conservative boundary conditions to a small but not limiting temperature increase in the storage pond, during transport and at final disposal.

This conclusion on activity and decay heat takes into account CEC studies completed in 1982 as well as actual results, and corrects the misleading conclusions drawn in the summary report of the CEC studies on "Control and safety of LWRs burning Pu fuel" [10]

2.3.2 Calculational methods

This section addresses in general the question of nuclear data and methods for application to Pu use in LWRs. The particular emphasis will be to comment on aspects which are of importance in dealing with multiple recycling of Pu in LWRs.

The situation of nuclear data

As a first comment, the nuclear data of all Pu isotopes should be taken from modern data libraries such as for example the European Joint Evaluated Files JEF-1 or JEF-2 [11], the US ENDF/B-VI [12] or Japanese JENDL-3 [13]. These data were assessed in many international discussions and finalised following these discussions. They have been checked carefully in applications covering both LWRs and fast reactors. These data have been available for some time from the OECD/NEA Data Bank. In case other data sets are used in calculations for Pu recycling as in the benchmarks (see Chapter 3), a comprehensive assessment of data differences is needed, which may subsequently help to explain any substantial differences in the final results and possibly indicate data which would need improvement (i.e., uncertainties' reduction).

Similarly detailed attention should be given to the generation of effective group constants, especially if only a few neutron energy groups are used in the calculation. Here again fuel isotopes are the most important ones. The effective group constants are generated mainly by using the code system NJOY [14], also available from the OECD/NEA Data Bank and RSIC.

As a special aspect, Pu-242 should be mentioned. In conventional LWR applications this isotope occurs at very low concentrations and needs no resonance self shielding treatment. But in multiple recycling schemes the concentration of Pu-242 is increased and resonance self shielding can no longer be neglected, as it will then have a significant influence on k-effective.

As regards determining the coolant void coefficient when a large part of the core is voided, it is in principle necessary that the voided zones be treated with a group constant set, generated with the harder weighting spectrum of the voided areas. This aspect is often neglected, and affects the reactivity change Δk_{void} calculated.

Another effect that has to be considered is the following: the different fissile isotopes, such as U-235 or Pu-239 have different fission spectra. Therefore in principle an average fission spectrum is needed, necessitating an iterative procedure in nuclear design calculations. K-infinity may change by up to 1% while the reaction rate ratio (i.e., U-235 fission to Pu-239 fission) can change up to 5%.

A point worth noting is that the use of appropriate models for neutron scattering on H_2O molecules is essential. The model used nowadays is that of Haywood [15]. With other simpler models (e.g., the heavy gas model), the thermal flank of the neutron spectrum in the thermal energy range is not properly described.

Another important aspect is the treatment of gadolinium as a heterogeneous absorber in the form of rodlets in the fuel bundle. Here the theoretical description is more laborious. The nuclear data used are sometimes uncertain and further improvement of the data should be sought.

As a last point, the spatial shielding via the Dancoff factor gives non-negligible effects, especially with regard to the question of how the fuel clad is taken into account, or indeed whether it is accounted for at all.

Many of the aspects mentioned above have been thoroughly studied in the context of investigations for the tight lattice Advanced Pressurized Light Water Reactor (APWR). The calculational methods have been checked against experimental results from the PROTEUS facility at PSI/Switzerland [16]. A very comprehensive discussion of data and methods is given in [17].

The determination of the space-energy distribution

The methods used are conventional, with transport methods such as S_n, collision probabilities, or Monte Carlo codes. These methods usually are very well tested. Recently nodal methods have attracted increased interest [18] for application in LWRs. They are also a very powerful tool because a reduction in computing time can be achieved. If those tools are used, then a detailed description of the fine flux reconstruction inside a calculational mesh has to be given [19].

2.3.3 MOX design issues

The neutronic design of Pu bearing fuel rods, fuel assemblies and cores has to take into account the conditions given by the established LWR technology. Changes in dimensions e.g., of fuel rods, rod cells, and fuel assemblies could cause incompatibility to existing PWR and BWR systems and could only be realised in the case of new reactor systems.

On the basis of elementary cells the changes in properties caused by the Pu content can easily and explicitly be discussed. The design of fuel assemblies reflects the need to ensure compatibility with UO_2 fuel assemblies as long as MOX and UO_2 fuel assemblies are in the same reactor at the same time. So the MOX fuel assembly design takes place preferably in a geometric model including more than one assembly. As the cores have to obey many safety related and economic boundary conditions as well, the fuel assembly design is not independent of the core design. The properties of respective cells cannot answer the question on global core properties which are mainly dependent on the amount of MOX in the core.

Plutonium-containing cells

When comparing Pu containing cells with similar UO_2 cells, for given dimensions of fuel pellets, canning and lattice pitch, it emerges that the Pu content causes a hardened thermal neutron spectrum (Figure 2.7) and a stronger under moderation as shown in Figure 2.8, where k-infinity is displayed as a function of fissile content and moderation ratio. The optimisation of the initial reactivity at a given Pu_{fiss} content of the fuel requires a widening up og the lattice, or, to the extent that this is forbidden for exixting reactors (especially so far PWRs), the changing of the rod/pellet dimensions as insertion of some water rods in the MOX lattice. With increasing moderation, however, the conversion ratio decreases. These two opposite effects favour unchanged MOX lattices as long as no other limiting conditions as e.g., a positive void effect at very high Pu contents, are reached.

MOX fuel assemblies

The neutronic design of MOX fuel assemblies has to allow for the effects caused by neighbouring fuel assemblies [4,5]. As long as UO_2 assemblies are in the core there must be provision made for the transition from the UO_2 neutronic spectrum to the spectrum in the MOX fuel rod areas. The design and optimisation of the Pu content in the different types of MOX rods aimed at avoiding too high power peaking in the area of spectrum transition is normally supported by calculations with a sufficient number of (thermal) neutron groups in a geometry including UO_2 and MOX rods. Such a geometry can be treated by two-dimensional transport calculations using pin cell multigroup cross-section data of spectral burnup calculations or directly with two-dimensional multigroup spectral calculations.

In the centre part of a MOX rod area, the plutonium content has to be high enough to provide the required reactivity [2,4]. It is generally advantageous to have rather large areas of this kind. This favours fuel assemblies with MOX rods only, where the fissile plutonium content must be lowered only in the outer rods in order to avoid the power peaking induced by the thermal flux of the surrounding uranium fuel.

If burnup equivalence is required for MOX to UO_2 fuel this has to be assured by core calculations for unchanged reload fractions and cycle length. A local power peaking form factor of 1.10 is obtained by using two to three plutonium content zones in the MOX fuel assemblies for PWRs.

Two design aspects must be balanced:

- *The power distribution within the MOX areas surrounded by uranium rods* must be flattened out by using a minimum of different fissile plutonium contents and adjusting the distribution of the respective rods;

- *The reactivity and burnup potential of the MOX fuel assembly* must be adjusted with respect to the uranium fuel assemblies via its average fissile plutonium content.

Therefore, in a PWR, design starts with a power distribution study in a macrocell model of a MOX fuel assembly surrounded by uranium fuel assemblies. For symmetrical fuel assemblies, the calculation can be done for one-eighth of a MOX fuel assembly with three-eighths of the uranium fuel assemblies building a triangle as shown in Figure 2.9.

In MOX fuel, there is an incentive to select the highest possible plutonium concentrations under a given power peaking limitation and to use only a minimum number of different plutonium contents. Thus, in the self-generated plutonium recycle mode, only one-third to one-fifth of all fuel rods in the reload and the core contain plutonium, depending strongly on the average core burnup, the plutonium composition and the carrier material.

Pin power reconstruction methods exist to include the rodwise results acquired during MOX-FA design into the core calculations with a small number of neutron groups (e.g., only 2) and coarse mesh representations (up to 1 box /FA).

The design of the MOX fuel assemblies under the above mentioned conditions is such that, in conjunction with unmodified reload strategies and an unchanged number of reload fuel assemblies (with or without U-Gd fuel assemblies), they achieve burnups comparable to uranium fuel assemblies and do not noticeably alter the length of the cycle. Early in life, this design approach produces slightly lower average linear heat generation rates than uranium fuel assemblies, together with a somewhat more wavy power distribution in the MOX-FAs throughout insertion.

The same considerations are valid for MOX-FAs for BWRs and PWRs, respectively [4]. Whereas in BWRs the MOX-FAs are rather weakly influenced by the surrounding FAs, they are strongly influenced by the water gaps between the FAs and void fraction inside FAs.

The extreme situation occurs when designing a 100% MOX-core. In this case there is no need to ensure compatibility with UO_2-FAs. Moreover, there are no constraints on the assembly geometry and the rod and cell geometry can be altered to optimise the core for Pu use. As long as no other limits are reached, one would use approximately the same moderation as in use in normal U-feed LWRs. For PWRs different Pu contents in the MOX-FAs are not necessary. However, in a 100% MOX BWR it is still essential, as a consequence of the water gaps, to have a Pu variation over the MOX-FA.

MOX-containing LWR cores

Given MOX-FA designs which obey the requirements of design and compatibility with UO_2-FAs, the core properties are determined by the core loading pattern and are generally perturbed to an extent which reflects the MOX content.

A prerequisite for plutonium recycling is the granting of a license for the use of MOX fuel assemblies in the reactor on the basis of given design requirements. Therefore, the technical feasibility is examined on the basis of realistic and enveloping designs. For this purpose, studies are carried out for different categories of requirements, as shown in Table 2.2 [4]. On this basis the validated limits of MOX fuel use are defined in the licensing procedure. Within these limits, licensing for individual cycles is then simply a matter of proving that existing analyses cover the case at hand.

Important cycle characteristics for various examples of PWR equilibrium cores with MOX fuel assemblies are listed in Table 2.3, [4]. The assessment of the core characteristics is considered with reference to the differences between MOX and uranium equilibrium cycles. With MOX fuel assemblies in the core, the more negative moderator temperature coefficient and the smaller boron worth are especially apparent. As regards the net control assembly worth for the stuck rod configuration at EOC in the hot-standby condition, data depend more on the loading scheme than on the fraction of MOX fuel assemblies in the core. Thus, the change from the traditional out-in reload pattern to a low-leakage one decreases the stuck rod worth enough to allow a MOX fraction of up to approximately 50% without the need for more control rods.

Table 2.4, [4], illustrates important characteristics of an equilibrium core with 31% MOX fuel assemblies for a 1300-MWe BWR. Core loadings have been investigated with up to approximately 50% MOX fuel and fissile plutonium concentrations providing sufficient reactivity to be equivalent to uranium fuel designed for an average discharge burnup of 45 MWd/kg. Especially in those cases of large amounts of fissile plutonium, it is important to mitigate the slightly less favourable neutronic characteristics of MOX fuel in comparison to uranium fuel, such as decreased control blade worth, burnable poison effectiveness, and increased void reactivity feedback.

Transient and accident analyses for LWR cores containing MOX fuel showed only small differences compared with those for pure uranium cores. For the rod drop accident as the limiting reactivity transient, the positive influence of the reduced control blade worth and the increased temperature and void reactivity feedback as well as the reduced number of delayed neutrons are important. Overall, no significant changes occur by the introduction of MOX fuel.

Under realistic operational conditions, the more negative void coefficient will even lead to a more favourable behaviour during that transient in comparison with a core without MOX fuel assemblies.

2.3.4 Verification of plutonium recycle cases

Measures to qualify the calculational methods to be used in design and modelling of MOX insertion in LWRs take various forms:

- Performing theoretical benchmarks, e. g., as performed by this working group,

- Critical experiments for realistic fuel configurations by various interested organisations,

- Special measurements and normal core follow for MOX-containing cores, including post irradiation experiments as e. g., rod to rod relative power and burnup distribution inside FA and isotopic composition measurements.

Because of the commercial implications, complete information on these verification and validation data is accessible only to the organisations directly involved. Only a survey from published results can be given in a report such as this.

Theoretical benchmarks

A CEC benchmark of 1977 on MOX pin cell burnup calculations for a PWR (SENA) and a BWR (Garigliano) showed k-infinity variations of up to about 10% for different codes [20].

Benchmarks related to tight lattice reactor concepts [21] covered different moderating ratios and voiding and suggest that practicable designs are feasible up to Pu contents of about $12 \pm 1\%$.

The WPPR has defined a new set of benchmarks covering different Pu compositions and higher Pu contents to be used in future in LWRs. Their improved results, compared to the mentioned CEC benchmark of 1977, are presented in detail in Chapter 3.

As the calculations are based at most on comparable nuclear data, an experimental verification would be of high interest. For better understanding the relevance of the calculations related to voiding, detailed calculations for reactors (e.g., three-dimensional realistic reactor voiding) could be of significance.

Critical experiments

In relation to early Pu recycling programmes several (clean) criticals have been measured, e.g., Battelle criticals [22] and Westinghouse criticals [23], (see Table 2.5). As they were carried out using low Pu content and also Pu of low Pu-240 content, they are no longer relevant to the present situation and to future validation requirements, except for the utilisation of military Pu.

If Pu with large fractions of the higher isotopes is used at Pu concentrations which are not too high, the basic cross sections are sufficiently qualified with the relevant calculation methods so that there is no need for further clean critical experiments. But the measurements carried out in connection with the design of an epithermal high conversion reactor (HCLWR) have relevance, especially in connection to the question of limits due to a positive void coefficient i.e., experiments in PROTEUS and EOLE [24].

The main questions related to starting large scale MOX use in LWRs that were answered by critical experiments performed since the early 70s were :

- The spectral transition from UO_2- to MOX-areas,

- The power distribution in the vicinity of UO_2/MOX-borders,

- The reactivity worth of control rods,

- The moderator temperature coefficient of such configurations,

- The interpretation of local activation measurements and fission chamber signals.

These programmes were carried out at the zero-power facilities VENUS, KRITZ, MINERVE and EOLE [5,24,25,26,27].

As early as the mid-60s, Belgonucléaire and the Belgian Nuclear Research Centre (SCK-CEN) started a close collaboration to perform R&D experimental work on the use of MOX fuels in LWR lattices. A full series of critical experiments were conducted in the VENUS critical facility with SS-clad fuel rods containing a moderate amount of relatively clean Pu (> 80% Pu-239, 17.4% Pu-240) (see Table 2.6); fission density distributions were particularly investigated around different lattice perturbations (water gap, absorbing rods, boundary MOX-UO_2, etc.).

In the early 80s, the same MOX rods were used in another series of experiments devoted to pressure vessel surveillance dosimetry, where computed and measured fission density distributions were shown to agree within a few percent.

At the end of the 80s, in the context of large-scale MOX utilisation programmes being considered in various countries and on the basis of their many years of expertise in the subject, Belgonucléaire and the SCK-CEN resumed their collaboration to implement the VENUS International Programmes (VIP); additional MOX rods were made available with up-to-date Pu amounts and isotopic vectors (Zr-clad, about 62% Pu-239, about 24% Pu-240), and also a number of UO_2-Gd_2O_3 poison rods (see Table 2.6). A programme was devoted to Pu recycle in PWRs and another one in BWRs. Basic measurements such as critical mass, axial bucklings, radial fission density distributions and detector responses, were performed in VENUS within mock-up configurations representing as close as possible current BWR or PWR assemblies; several core loadings were investigated in detail. Typical ones for both series are illustrated on Figures 2.10 and 2.11, where the measurements are also compared with the results of Belgonucléaire computations; (C-E)/E values derived from SCK-CEN calculations have been presented at the Tel Aviv Conference [28].

Another VENUS programme is under way to investigate the reactivity effects of moderator voiding in MOX zones with different (high) Pu contents; critical masses, axial bucklings and fission density traverses are measured with and without a central void in the MOX zone, and corresponding calculations are being performed by all partners.

31

The KRITZ facility is a flexible device for reactor physics measurements on water moderated cores of full length LWR fuel rods at temperatures up to 245°C. The programmes KRITZ-2 to KRITZ-4 (as listed in Table 2.7) in the early 70's were devoted to Pu recycling in BWRs and PWRs.

KRITZ-3, conducted for Siemens with fuel that was later on inserted at KWO (NPP Obrigheim), refers to various PWR cores. The fuel region in each of these cores is approximately cylindrical with a radius of about 30 cm. The fuel rods consist of UO_2 with a U-235 enrichment of about 3%. In two of the cores, the UO_2 rods in the central assembly were replaced by $U(nat)O_2$, PuO_2 rods, some of which had 2% fissile Pu and some about 3%. The boron content was between 400 and 1300 ppm. Some more details are given in Table 2.8.

The EPICURE programme was agreed in 1987 within a framework of a collaboration between CEA, EDF and Framatome, in order to build an exhaustive experimental database related to Pu recycling in PWRs and consequently to enable the uncertainties associated with MOX fuelling to be reduced [25].

By means of the chosen experimental strategy, most of the main physical phenomena of such reactors have been measured. Clean core characterisations enable the qualification of both nuclear data libraries and cell codes; mock-up cores allow the investigation of power distributions by means of fission rate distribution measurements, the qualification of calculations of power maps within MOX-UO_2 assemblies and at the interfaces and the interpretation of in-core chamber measurements in operating reactors [29].

K-effective calculations performed with the new CEA-93 library (JEF-2.2 evaluations) and the APOLLO-2 code show an overestimation by about 300 pcm for the UO_2 (3.7% U-235 enrichment) lattice and an underestimation of about 750 pcm for the $UPuO_2$ clean core (7% Pu).

Calculations of the mock-up configurations are in very good agreement with the experimental results (1 standard uncertainty margin: +/-4%) and CEA considers that the power distributions within the subassemblies are given to within 1.5% (1σ) with the improved calculational scheme (S_n transport theory - 99 groups). Nevertheless, the standard calculational scheme (2 group diffusion theory) gives 4% to 6% discrepancies at the MOX-UO_2 interface [30].

Furthermore, specific experimental measurements have been devoted to study the local voiding effect in both clean UO_2 and clean $UPuO_2$ cores:

- *2-D voided configurations* (30%, 50% and 100% of void): reactivity effect and power distribution measurements in both UO_2 and MOX clean lattices;

- *3-D voided configurations* (simulation of a "bubble") were planned in 1994 (7% and 11% Pu rods).

The last phase of the EPICURE programme will be devoted to temperature coefficient measurements in both UO_2 and MOX clean lattices in order to evaluate the uncertainties related to Doppler coefficient and moderator temperature effects. The experimental programme consists of a comparison between cold and "hot" (about 85°C) clean UO_2 and MOX configurations with moderation ratios (moderator volume / fuel volume) of 1.2 and 1.5. This will enable water density effects to be separated from the temperature effects (Doppler broadening effect and thermal scattering effect).

Typical MINERVE results and an EPICURE core configuration are given in Figure 2.15. The given EPICURE MOX arrangement is comparable for example to KRITZ-3 Pu-WH1.

As reported by the conducting organisations the experiments had been very valuable in validating the calculational procedures. It was stated, that a comparable agreement was found as for measured comparable UO_2 arrangements.

For several years the PROTEUS facility in Switzerland was used for the study of the physics properties of HCLWR lattices. The Phase II series of experiments [16, 31] comprised the investigation of very tight and wider lattices fuelled with MOX. In order to provide a more thorough understanding of the physics of MOX fuelled and well moderated lattices, a wide lattice with moderator-to-fuel ratio typical for a more conventional LWR was investigated. The high enrichment fuel originally manufactured for HCLWR research was also used in this test configuration.

A large variety of integral experiments, among them the measurement of reaction rate ratios and k-infinity was performed in a central test zone of 0.5 m diameter and 0.84 m height. All test zone configurations consisted of stainless steel clad fuel pins with an outer diameter of 9.57 mm arranged in triangular lattices. In the tight HCLWR lattices the pitch-to-diameter (p/d) ratio was 1.12. In the wider HCLWR lattices it was increased to 1.26. The effective moderator-to-fuel volume ratio (V_m/V_f) of these lattices was 0.48 and 0.95, respectively.

The well moderated MOX lattice was created from the wider H_2O-moderated HCLWR lattice by removing every third fuel pin in a central region of about 0.24 m diameter. The effective moderator-to-fuel volume ratio of this lattice was thereby 2.07. The comparison of parameters measured and calculated provide a valuable basis for validating calculations of high enrichment MOX fuel in LWRs.

In France, within the framework of a co-operation between the French utility EDF, the reactor designer Framatome and the French Atomic Commission CEA, an exhaustive experimental programme was undertaken in the centre of the EOLE facility at the Cadarache Research Centre in order to measure the main fundamental neutronic parameters involved in tight lattice HCLWRs design calculations [32,33].

Two types of lattices have been investigated:

• A very tight one, called ERASME/S (Mod. Ratio = 0.5) and,

• A more realistic one, ERASME/R (Mod. Ratio = 0.9) in order to be representative for Framatome's HCLWR projects.

Calculations are in good agreement with the experimental values [34].

Void configurations have also been investigated in both lattices. The reactivity effect was negative because of the very large effect of axial leakage. Calculations are in good agreement with the experimental results.

In parallel, in two similar undermoderated lattices placed in the zero power MINERVE facility and in the 8 MWth MELUSINE core, neutronic parameters related to the evolution with irradiation (fission products reactivity effect and capture cross-sections of the main heavy isotopes) have been measured [35].

Calculational results show some discrepancies on capture cross-sections but the fission product reactivity effect seems to be well estimated [36].

Ongoing experiments are devoted to the necessary changes in MOX designs with higher Pu content and for Pu with degraded content of thermal fissile isotopes. In this area the theoretical results need further experimental verification.

Experience with MOX-containing LWR cores

After the early Pu recycling programmes depicted in Table 2.9, the initial commercial MOX insertion in LWRs was concentrated in PWRs. Figure 2.13 gives a survey on the German programme (by Siemens) including the starting phase at KWO (NPP Obrigheim) and the Swiss Plant Beznau-2 for the time interval 1972-1993.

In parallel, 4 test-MOX-FAs had been inserted (by Westinghouse) in 1978-81 in Beznau-1 and further 36 MOX-FAs had been loaded in this reactor in the time span of 1988-1990.

The French programme for the EDF 900-MW PWRs starting with a first reload including 16 MOX-FAs in 1987 at St. Laurent B1 is given in Figure 2.14 (and including 6 plants presently). This programme includes the insertion of 16 further MOX-FAs in SENA during 1987-1991.

Ten additional 900-MW PWRs of EDF are licensed to recycle the Pu produced by the reprocessing of spent fuel, depending essentially on the availability of MOX fabrication capacity.

In Germany three 18x18 KONVOI PWRs are licensed for the irradiation of up to 50% MOX-FAs; several BWRs have requested MOX insertion licences [4].

The neutron physics experience acquired in these reactors is based on :

- Start-up measurements,

- In-service cycle monitoring, and,

- Specific measurements.

The reliability of the design methods is confirmed by measurements of cycle length, boron concentrations, reactivity coefficients (such as for coolant temperature and boron worth), control rod worth, and power density distribution.

In addition to the measurements assuring the shutdown margin, special measurements were conducted to investigate the worth of control assemblies inserted in MOX-FAs. Starting e.g., from the KRITZ-3 measurements for the insertion of control rods into a UO_2- and MOX-FA-arrangement for fresh fuel as a function of the temperature, further measurements were conducted to answer the question for partly burnt MOX-FAs directly in MOX-fuelled PWRs via boron compensation and bank exchange techniques.

No significant increase in the deviations between measurement and calculation were found with increasing MOX content of succeeding cycles if modern data sets and calculational tools were used. Table 2.10 gives examples taken from the French measurements on critical boron concentration, isothermal coefficient and several control bank worth values [5,37].

Regarding the validation of power distributions [4,5], two examples are given in Figure 2.15 and Figure 2.16. They cover cycle 5 of the German NPP Grohnde in full low-leakage loading measured by the aero ball system and the EDF power plant St-Laurent-1 with the content of 3x16 MOX-FAs (cycle 7), measured by movable detectors.

A special aspect in the field of validation is that of isotopic composition measurements. Such measurements have been done on fuel rods extracted from MOX-FAs after 1 to 4 irradiation periods for first and second generation MOX fuel rodlets and for reprocessing batches of MOX-FAs irradiated in German PWRs. Included in these were burnup measurements. The calculational results on the basis of power histories give the opportunity to control the building and transmission rates covering Pu isotopes and higher actinides, too. The actual cross-section sets are valuable today as they cover the present day Pu isotopics to satisfactory accuracy.

The task of validation is ongoing, as new MOX-FA designs, higher Pu contents and higher amounts of MOX-FAs in the cores are introduced. Looking forward, the introduction of MOX insertion in BWRs will also be accompanied by validation measurements.

2.4 Mechanical and chemical properties of MOX fuel and MOX fuel rods

2.4.1 General aspects [2]

Considerable effort has been devoted to determine the physical properties of mixed oxide fuel for fast reactors, which has about 20 per cent plutonium content. These studies have shown that several of these properties are poorer than those of UO_2. However, these data are overly conservative for application to MOX fuel for water reactors, which has a plutonium content of 4 to 10%. For such fuel, the performance and safety-related characteristics can be summarised as follows:

- Thermal conductivity is influenced by stoichiometry, porosity, plutonium content and irradiation. Correlations have been developed to incorporate the correct input in the fuel modelling codes;

- The melting point of stoichiometric MOX (at 5% plutonium) is about 20°C below that for UO_2. Within practical limits, hypostoichiometry does not have an effect on the melting temperature;

- Nuclear self-shielding is more pronounced in MOX fuel than in UO_2. Hence more heat is generated at the periphery in MOX fuels, mitigating the effects of the poorer thermal properties of MOX fuel;

- Thermal expansion is about 1% higher than for UO_2;

- MOX fuel exhibits better creep properties than UO_2 fuel. This is the most likely reason for its better pellet clad interaction behaviour;

- The homogeneity of the plutonium distribution in the uranium matrix of the pellets depends on the fabrication route. It directly affects fission gas release, densification and swelling, thermal limitation under reactivity excursions and the capability of the fuel to be reprocessed.

Experience has shown that a properly founded design technology combined with adequate manufacturing techniques is adequate to provide for the engineering and the licensing of MOX fuel without undue conservatism.

The chemical properties of Pu are only relevant at the reprocessing stage of the fuel cycle, where the solubility of the oxide and the treatment of Pu-nitrate and its conversion to Pu-oxide is important.

2.4.2 MOX fabrication [2,38,39,40,41]

MOX fabrication is in principle comparable to UO_2 fabrication. The differences are caused by the radioactivity and radiotoxicity of Pu, which is used only in the form of PuO_2. Commercial PuO_2 is extremely radioactive. It is essentially an alpha emitter, although it is also a producer of neutrons, X-rays, gamma rays and beta particles.

Plutonium's alpha and beta emissions and its toxicity require it to be handled in gas tight glove boxes provided with large plexiglas windows and suitable gloves.

The gamma activity builds up continuously after the last purification step of the PuO_2 at the reprocessing plant. The main source is the decay of Pu-241 into Am-241. This major contributor is relatively easy to shield, since the radiation is at a low energy level. The only problem stems essentially from dust deposition on equipment and on the internal surfaces of glove boxes.

Most of the high energy radiation arises from the decay of Pu-236 into Bi-212 and Tl-208 causing hard gamma radiation. This source is, however, much smaller than the one from Am-241 and builds up at a rate five times slower.

Neutron activity increases slightly with time and depends mainly on the isotopic composition of the plutonium.

The heat generated by the alpha activity (especially from Pu-238) is somewhat higher during storage and a gradual degradation of the ease of fabrication results if PuO_2 is stored over long periods. The problems of gamma and neutron activity and heat generation increase with increasing burnup of the irradiated fuel from which the Pu has been separated. With the trend toward very high discharge burnups, the MOX fabrication plant must be heavily shielded with automatic process equipment which minimises operator involvement both for operations and maintenance to control the dose to the plant operators.

Plutonium is usually transferred from the reprocessor to the fuel manufacturer as PuO_2 powder. If plutonium oxide is stored for too long before proceeding with fuel fabrication, an extra processing step,

Am-241 separation, may be required unless the MOX fabrication plant has been designed with the necessary additional shielding. It consists of dissolution, purification and reconversion to PuO_2.

Proper shielding has to be taken into consideration at all stages of the MOX fuel cycle; although it is usually not a limiting feature for LWR fuel.

Until 1981 MOX fabrication for use in LWRs started from sinterable UO_2 powder of good flowability. In most cases, PuO_2 was delivered as powder calcined from precipitated oxalate with very fine particles or was prepared directly from plutonium/nitrate solutions. Mechanical blending of such powders gave a homogeneity with PuO_2 particle sizes which prevent hot spots in the MOX. Such fuel met the requirements of operation and exhibited excellent irradiation behaviour. Results from reprocessing of such fuel showed a residual insolubility in pure nitric acid.

Therefore new powder preparation routes were developed to meet the new solubility specifications. The OCOM and MIMAS methods (see Figure 2.17) avoid pure PuO_2 particles by co-milling or micronising a master-mix of about 30% Pu, which is then blended with pure UO_2 at the final Pu content to achieve complete homogenisation of the mixture. The COCA process avoids the master-mix and uses the final Pu content at the milling step. The Short Binderless Route (SBR) uses high energy co-milling of PuO_2 and Integrated Dry Route (IDR) UO_2 at the final Pu content to achieve complete homogenisation of the mixture (see Figure 2.17). Alternative routes have been developed which produce mixed oxide powders directly from a blend of Pu and U nitrates, e.g., the German AuPuC process (Figure 2.17) or the Russian GRANULAT process. These routes are not presently used on the industrial scale, however.

An important aspect of MOX fabrication is the treatment of isotopic variations in the source Pu. Differences in the isotopic makeup of the Pu would cause unacceptable variations in the physics performance of MOX assemblies unless special measures were in place. There are several different approaches to this question. The approach adopted in the Hanau plant involves characterising the isotopic makeup of various PuO_2 lots and blending precisely the same quantities from each lot to make up all of the MOX powder required for a particular batch of MOX assemblies. This ensures that all the MOX assemblies have the same Pu isotopic mix. The approach adopted in the MELOX plant is similar, but involves the solution of a set of simultaneous equations to determine the appropriate blending ratios to attain the desired Pu isotopic makeup in the final product. This has the advantage that fewer PuO_2 lots need to be available at any one time. The solution that has been adopted for the Sellafield MOX Plant (SMP) involves pairing batches of PuO_2 to obtain equivalent nuclear performance. In this instance, the isotopic makeup of the final product is allowed to vary within carefully defined limits.

It is not appropriate in this report to dwell on the details of fuel pelleting and fuel rod and assembly fabrication, but some important points are worth noting :

- An appropriate heat treatment may be necessary for assuring the requisite low hydrogen contents of the fuel pellets;

- In the absence of forced cooling the heat generation in powders places restrictions on the batch hold times before pressing since the heat destroys the lubricant;

- Dust buildup on the inside of the glove boxes challenges the plant shielding (since it is additional to the bulk source) and increases maintenance doses;

- The fuel rod fabricationalso has to be done within glove boxes (Figure 2.18);

- Particular issues in fuel rod fabrication are contamination of the weld area and confirmation of the correct Pu content;

- The use of multiple Pu content levels in the assembly places constraints on manufacture. This is especially important for BWRs.

2.4.3 In-core behaviour of MOX fuel

The design of MOX fuel rods follows the procedure for UO_2 mechanical fuel and rod design. The small perturbation in thermo-mechanical properties caused by the content of PuO_2 as mentioned in section 2.4.1 are included. The less homogeneous structure of the fuel by embedded master mix particles in the UO_2 matrix is of no relevance as long as the MOX particles are small enough to avoid local hot spots in the inner part of the pellet as well as on its surface.

The structure determines the fission gas release and the dimensional behaviour during irradiation. As the power histories of MOX fuel rods tend to have a higher power at high burnup due to the less steep fall-off in reactivity with burnup compared with UO_2 fuel rods, somewhat higher fission gas releases are calculated. The rod design has to take care of this effect by proper design of the plenum in the rods.

The models used for the rod design have to be qualified by appropriate experimental data.

The irradiation behaviour of MOX fuel has been investigated in detail by surveillance of many MOX fuel assemblies in the different spent-fuel pools. In addition, irradiation programmes with pathfinder MOX fuel rods in special carrier fuel assemblies and with special MOX test rods in test rigs in selected nuclear power plants as well as in test reactors were performed followed by regular post-irradiation examinations (PIEs) in the spent-fuel pool and in hot cells [42,43].

Investigations of MOX fuel rods show that the overall rod dimensional behaviour is similar to UO_2 rods. This is because identical cladding tubes were used for the rods examined and the MOX fuel density behaviour was found to be similar to UO_2 fuel up to burnups of 50 MWd/kg.

Understanding the density behaviour of MOX fuel requires consideration of its structure. On a microscopic scale, the fuel structure appears heterogeneous with MOX agglomerates uniformly distributed. Contributions of the high local burnup in the MOX (master mix) particles, the matrix swelling rate related to the matrix burnup and the development of the porosity sum up to the global dimensional behaviour of the MOX fuel.

Density measurements show a similar dimensional characteristics of OCOM (as well as comparable MIMAS- and COCA-type MOX fuels) and AUPuC types of fuel. This results from the superposition of two-dimensional processes: the delayed densification and swelling of the matrix due to the low burnup and the swelling of the MOX agglomerates.

An analysis of the fission gas release shows that temperature has a strong influence. The highest gas release always occurred at the highest temperatures. Under steady-state conditions, this was concluded from the power history of the MOX fuel rods. Transient test conditions are better suited to quantitatively investigate the effects. MOX and UO_2 fuel showed similar fission gas release at similar temperatures. This can be understood from the fact that the fission gas is always released via the UO_2 matrix. The remaining gas in the bubbles of the MOX agglomerates seems to be concealed, even at intermediate temperatures where the agglomerates are characterised by large bubbles within the agglomerates. Only at high temperatures, when release channels in the UO_2 matrix are formed, is the fission gas able to leave the fuel via these channels.

Transient-tested MOX fuel shows a dimensional behaviour comparable to that of UO_2 fuel. The transient fission gas release was also found to be similar to that in UO_2 fuel operated at the same temperature.

The tight enclosure of MOX agglomerates by the UO_2 matrix and the implantation of fission products and fission gases into the UO_2 matrix also prevent the instantaneous release of the fission products into the primary coolant in case of fuel rod defects and are considered to be the reason for the similarity in defect behaviour of UO_2 and MOX fuel.

In conclusion, a comparison of MOX and UO_2 fuel shows that both types of fuel, in spite of different structure and hence local burnup, have similar dimensional and fission gas release behaviour. Therefore, it is justified from a technical point of view to also use similar models for design calculations.

Ongoing testing of MOX fuel is needed as a consequence of the following reasons:

- Small changes in pellet density and diameter as well as canning dimensions and material properties influence the burnup behaviour;

- Planned changes in the MOX fabrication technology are to be investigated;

- Higher Pu contents are to be included at actual Pu compositions;

- An increase in burnup is planned.

2.4.4 Reprocessing aspects

In the reprocessing of MOX fuel, good solubility of plutonium in pure nitric acid is important. The actual fuel fabrication processes (AUPuC, OCOM, MIMAS, COCA) already warrant this property for the as-fabricated MOX condition (> 99%) [43].

To determine the solubility behaviour of plutonium from irradiated MOX fuel rods, fuel samples were examined in Germany from one-, two-, three- and four-cycle AUPuC and OCOM MOX fuel rods, respectively. The content of fissile plutonium was approximately 3.2% and the burnup of the samples was in the range of 3 to 46 MWd/kg. The solubility tests performed in pure nitric acid yielded high plutonium solubilities (> 99,8%) comparable with that of UO_2 fuel.

Similar experiments were conducted in France. They showed that irradiation in LWR reactors led to the eradication of most of the homogeneity flaws of the plutonium in the tested MOX fuels. The final solubility of plutonium was consequently very good (> 99.97%) and had no effects on the process conditions.

The feasibility of reprocessing MOX fuel as manufactured today was demonstrated on semi-industrial scale at APM (2.1 t in 1992), and then on the industrial scale in the Cogema UP2 plant of La Hague (4.7 t in 1992) in conditions similar to those employed to reprocess UO_2 fuel.

In total 16 MOX-FAs fabricated by OCOM and AUPuC processes burnt in four different PWR plants in Germany have been reprocessed under commercial contracts by Cogema. The analyses performed during the reprocessing operations confirm the good solubility of the plutonium in the operating conditions (solubility better than 99.95% with respect to the plutonium content before irradiation).

The technological options selected by Cogema for reprocessing MOX fuel in UP2-800 were accordingly confirmed. This industrial-scale reprocessing of irradiated MOX fuels opens the way to multiple recycling of plutonium in reactors.

2.5 Summary and conclusions

This chapter deals with the neutron physics aspects of Pu recycling in PWRs and BWRs. In addition, some engineering aspects such as the mechanical and chemical properties of MOX fuel are also discussed. Calculational results are compared with experimental findings from critical experiments, from experiments performed in commercial power reactors and recycling experience. The agreement for first pass recycling is fully satisfactory. Multiple recycling is discussed in the context of a new benchmark. The results show an improvement over the first such exercise of the European Community of several years ago, but are not fully satisfactory yet. Calculations on the basis of new nuclear data sets show improved results (see Chapter 3 for details). Taking account of the more recent nuclear data sets and more sophisticated calculational methods only, the deviations in k-infinity are less than 1% at zero irradiation and about 1.5% at a burnup of 50 MWd/kg.

Some of the principal conclusions – essential for multiple recycling – are as follows:

1. Regarding nuclear design calculations, modern methods incorporating rigorous resonance self-shielding and modern nuclear data libraries such as JEF-2, ENDF/B VI or JENDL-3 are essential. Some data improvement, e.g., higher plutonium isotope data in the resonance region, etc., might be needed;

2. Appropriate and well tested calculation methods are widely available and should be used;

3. The interaction of neighbouring UO_2 and MOX elements should be investigated in more detail to obtain a clearer understanding of the results observed;

4. The optimal control of high burnup LWR cores (which may include burnable poisons) should be examined further;

5. Additionally, questions of material damage in the case of high burnup need to be clarified as well as other engineering constraints;

6. Sophisticated (three-dimensional) core calculations are recommended to study the void effect and could also be the object of a future benchmark exercise, possibly in relation with the analysis of an experiment to be made available to the international community;

7. At present, it appears that plutonium recycling in high burnup LWR cores can be performed twice without modifying current LWR designs;

8. Future experimental verification related to maximum plutonium content in the case of degraded plutonium isotopic composition are needed in clean lattice configurations with different moderator-to-fuel ratios and for possible void experiments with different leakage components. A co-ordinated international effort in this field would be highly beneficial;

9. For unchanged lattices the limit on Pu content is in the range $12 \pm 1\%$;

10. Changes in the lattice in the sense of higher moderation can be foreseen to minimise the buildup of higher actinides and to increase the limits on Pu content;

11. Multiple recycling of plutonium with high burnup (e.g., 50 MWd/kg) can have limitations due to considerations such as the buildup of Pu-238 and Pu-242 or the existence of positive reactivity feedback effects on complete coolant voiding at high Pu contents or the increase of the buildup of higher actinides (Am, Cm, etc.). A specific future benchmark in this field could help in obtaining an international consensus on these limitations.

References

[1]   G. Schlosser, R. Manzel: Siemens Forsch.- u. Entwickl.-Ber. 8, 108 (1979).

[2]   H. Bairiot: Nuclear Engineering International 29, 27 (1984).

[3]   G. J. Schlosser, S. Winnik: IAEA-SM-294/33 (1987).

[4]   G. J. Schlosser, W.-D. Krebs, P. Urban: Nuclear Technology 102, 54 (1993).

[5]   M. Rome, M. Salvatores, J. Mondot, M. Le Bars: Nuclear Technology 94, 87 (1991).

[6]   I. K. Levina, V.V. Saprykin, A.G. Morozov: "The Safety Criteria and WWER Core Modification for Weapon Plutonium Utilisation", NATO Workshop Obninsk, 17-19 Oct. 1994, to be published.

[7]   Jahrestagung Kerntechnik '92, Fachsitzung "Wie ist die Entsorgung deutscher Kernkraftwerke in Zukunft sichergestellt?" B. Lenail, p. 61 and W.-D. Krebs, P. Schmiedel, p. 73.

[8]   Final GESMO, NUREG-0002 (1976).

[9]   M. Taube: "Plutonium - A General Survey", Verlag Chemie 1974.

[10]   J. Basselier, A. Renard, "Analysis and Synthesis of the Theoretical Studies Performed on the Control and Safety of Lwr's Burning Plutonium Fuel", EUR 8118 EN (1982).

[11]   J. L. Rowlands, N. Tubbs, JEF-1:" The Joint Evaluated File: A New Nuclear Data Library for reactor Calculations", Proc. of Int. Conf. on "Nuclear Data for Basic and Applied Science", Santa Fe, New Mexico, 13-17 May 1985, vol. 2, p. 1493 (1985)

        C. Nordborg, M. Salvatores: JEF-2: "Status of the JEF Evaluated Data Library", Proc. of Int.Conf. on "Nuclear Data for Science and Technology", Gatlinburg, Tennessee, 9-13 May 1994, vol. 2, p. 680 (1994).

[12]   R. W. Roussin, P. G. Young, R. McKnight: "Current Status of ENDF/B-VI", Proc. of Int. Conf. on "Nuclear Data for Science and Technology, Gatlinburg, Tennessee, 9-13 May 1994, vol. 2, p. 692 (1994).

[13]   Y. Kikuchi: "JENDL-3 Revision 2 – JENDL 3-2", Proc. of Int. Conf. on "Nuclear Data for Science and Technology, Gatlinburg, Tennessee, 9-13 May 1994, vol. 2, p. 685 (1994). .

[14]   R. E. MacFarlane: "The NJOY Nuclear Data Processing System", Version 91, LA-12740-M.

[15] B. C. Haywood, D. I. Page: "The Harwell Scattering Law Programme: Frequency Distributions of Moderators", AERE-R-5778 (1968).

[16] R. Chawla, R. Böhme, J. Alfonso et.al.: Kerntechnik 57, 14(1992).

[17] C.H.M. Broeders: KfK-Berricht 5072 (1992).

[18] M.R. Wagner: "Nuclear Science and Engineering" 103, 377 (1989).

[19] C. Cavarec, J. F. Perron, D. Verwaerde, J. P. West: "Benchmark Calculations of Power Distributions within Assemblies", NEA/NSC/DOC (94) 28.

[20] P. Reuss, J. P. Fischer, M. J. Basiuk: "Synthèse des calculs 'benchmark' sur le recyclage du plutonium dans les réacteurs à eau ordinaire", CEC/CEA Contract 0187612RPUF (1978).

[21] Y. Ishiguro, H. Akie, H. Takano: "Summary of NEACRP Burnup Benchmark Calculations for High conversion Light Water Reactor Lattices", Proc. Int. Reactor Physics Conf., Jackson Hole, Wyoming, Vol. III, 97 (1988).

[22] UNC-5211.

[23] R. D. Leamer, WCAP-3726-1 (1967).

[24] IAEA-TECDOC-638: "Technical Aspects of High Converter Reactors", Nuremberg, 26-29 March 1990, (1992, IAEA Vienna).

[25] J. Mondot, J. C. Gauthier, P. Chaucheprat, J. P. Chauvin, C. Garzenne, J. C. Lefebvre, A. Vallee: "PHYSOR'90", VI.53 (1990).

[26] A. Charlier, J. Basselier, L. Leenders: Proc. Int. Conf. on "The Physics of Reactor – PHYSOR'90", VI.65 (1990).

[27] P. F. Rose, "Proceedings: Thermal-Reactor Benchmark Calculations, Techniques, Results, and Applications", Upton, NY, 17-18 May 1982, EPRI NP-2855, RP975-1 (1983).

[28] G. Minsart, P. D'Hondt, Kl. Van der Meer: "Reactor Physics and Reactor Computations", Tel Aviv, 467 (1994).

[29] J. P. Chauvin, G. Granget, M. Martini et al.: Proc. Top. Meet. on "Advances in Reactor Physics", Vol. 2, 374 (1992).

[30] P. Fougeras, S. Cathalau, J. Mondot, P. Klenov: Proc. Conf. on "Reactor Physics Faces the 21st Century", Knoxville, Vol. 3,113(1994).

[31] R. Böhme, H.-D. Berger, R. Chawla et al.: "Reactor Physics and Reactor Computations", Tel Aviv, 223 (1994).

[32] A. Santamarina, L. Martin-Deidier, S. Cathalau et al.: Top. Meet. on "Reactor Physics", Saratoga Springs, Sept. 1986.

[33]   A. Santamarina, L. Martin-Deidier, S. Cathalau et al.: Top. Meet. on "Advances in Reactor Physics, Mathematical and Computation", Paris, (1987).

[34]   S. Cathalau: Proc. Top. Meet. on "Advances in Reactor Physics", Charleston, Vol. 2, 385(1992).

[35]   M. Darrouzet, J. Bergeron et al.: Nucl. Techn. 80/2, 269 (1988).

[36]   P. Chaucheprat, J. Mondot et al.: Top. Meeting on "Reactor Physics", Jackson Hole, (1988).

[37]   J. C. Barral, C. Hervouet, P. Larderet, J. C. Lefebvre, A. Vassalo, M. Lam-Hime, M. A. Bergeot: "PHYSOR'90", VI.34.

[38]   H. Röpenack, F. U. Schlemmer, G. J. Schlosser: Nuclear Technology 77, 175(1987).

[39]   H. Bairiot, G. Lebastard, J. F. Marin, F. Motte: Nuclear Europe 5/1985, p. 25.

[40]   J. M. Leblanc: Nuclear Europe 2/1984, p. 33.

[41]   J. Krellmann: Nuclear Technology 102, 18 (1993).

[42]   D. Haas: Nuclear Engineering International Feb. 1987, p. 35.

[43]   W. Goll, H.-P. Fuchs, R. Manzel, F. U. Schlemmer: Nuclear Technology 102, 29 (1993).

Mass [g/kg HM initial]

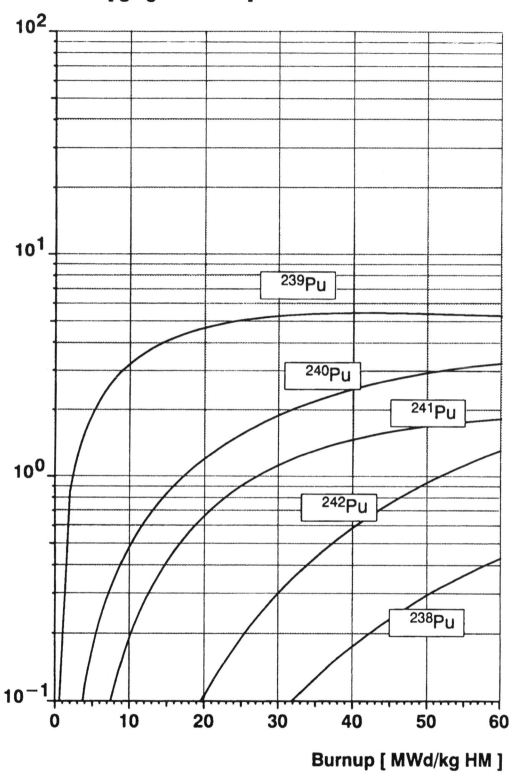

Figure 2.1 Buildup of Pu isotopes in a UO$_2$-FA of 4.0 w/o initial enrichment

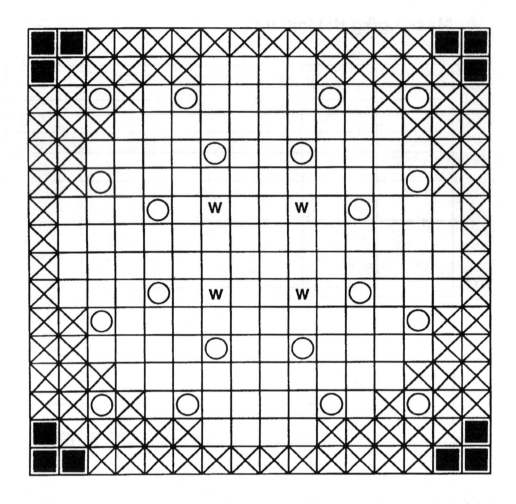

| Type of rod | Content of fissile material in w/o | | Number of rods |
	fissile Plutonium (Pu_{fiss}/heavy metal)	U235 ($U235/U_{tot}$)	
■	2.0	0.25	12
⊠	2.8	0.25	92
□	4.1	0.25	128
◯	– guide tube	–	20
W	– water rod	–	4

Figure 2.2 MOX-FA in use in German PWRs of 1300 MWe

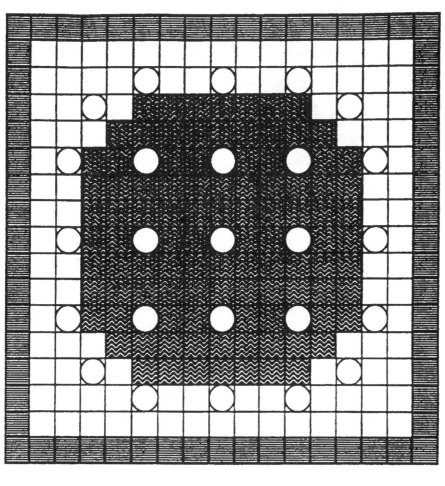

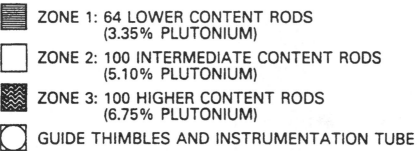

ZONE 1: 64 LOWER CONTENT RODS
(3.35% PLUTONIUM)

ZONE 2: 100 INTERMEDIATE CONTENT RODS
(5.10% PLUTONIUM)

ZONE 3: 100 HIGHER CONTENT RODS
(6.75% PLUTONIUM)

GUIDE THIMBLES AND INSTRUMENTATION TUBE

Figure 2.3 **MOX-FA in use in French PWRs of 900 MWe**

Control Rod Position

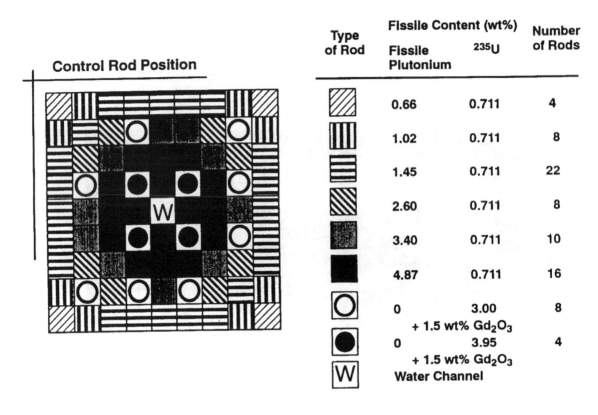

Type of Rod	Fissile Content (wt%)		Number of Rods
	Fissile Plutonium	^{235}U	
	0.66	0.711	4
	1.02	0.711	8
	1.45	0.711	22
	2.60	0.711	8
	3.40	0.711	10
	4.87	0.711	16
	0	3.00 + 1.5 wt% Gd_2O_3	8
	0	3.95 + 1.5 wt% Gd_2O_3	4
W	Water Channel		

Figure 2.4 Existing MOX-FA for BWR Gundremmingen B/C

Control Rod Position

Type of Rod	Fissile Content (wt%)		Number of Rods
	Fissile Plutonium	^{235}U	
	1.59	0.20	4
	2.58	0.20	8
	3.69	0.20	24
	5.33	0.20	24
	0	3.95 + 1.5 wt% Gd_2O_3	12
W	Water Channel		

Figure 2.5 MOX-FA proposed for German BWRs

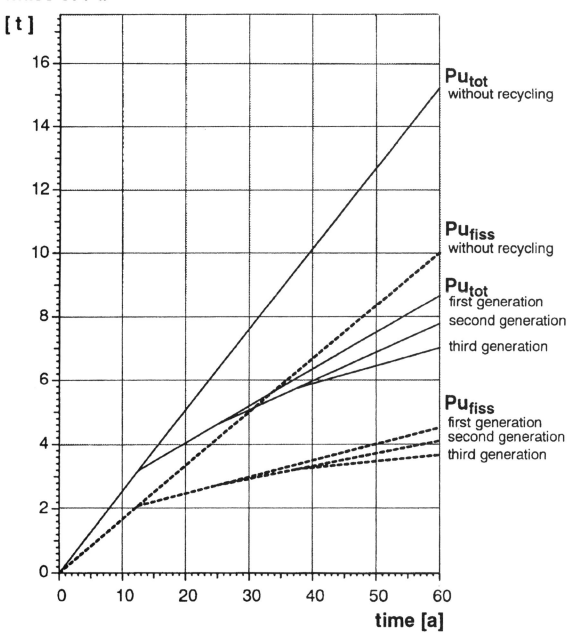

accumulated
mass of Pu

[t]

Pu_{tot}
without recycling

Pu_{fiss}
without recycling

Pu_{tot}
first generation

second generation

third generation

Pu_{fiss}
first generation
second generation
third generation

time [a]

Figure 2.6 Evolution with Pu multirecycling
PWR of 1300 MWe, 45 MWd/kg, recycling delay 12.5 years

49

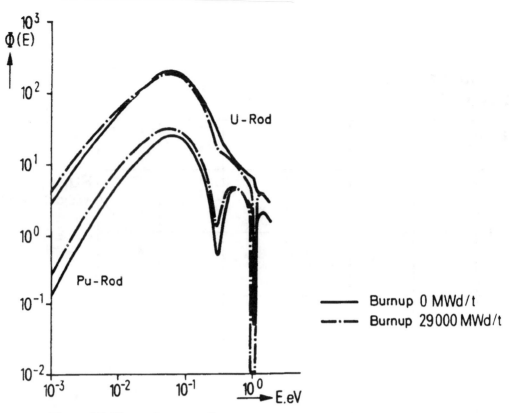

Figure 2.7 **Thermal neutron flux spectrum in UO₂ and MOX PWR fuel cells**

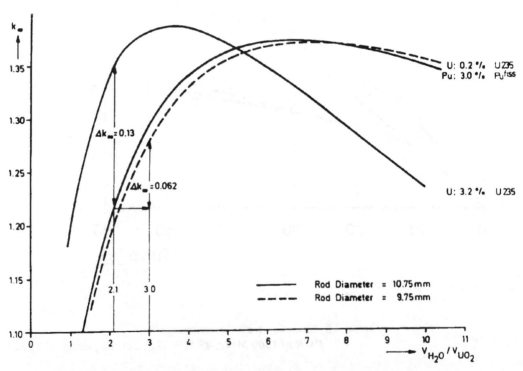

Figure 2.8 **k-infinity of a PWR vs. moderating ratio at operating temperature**

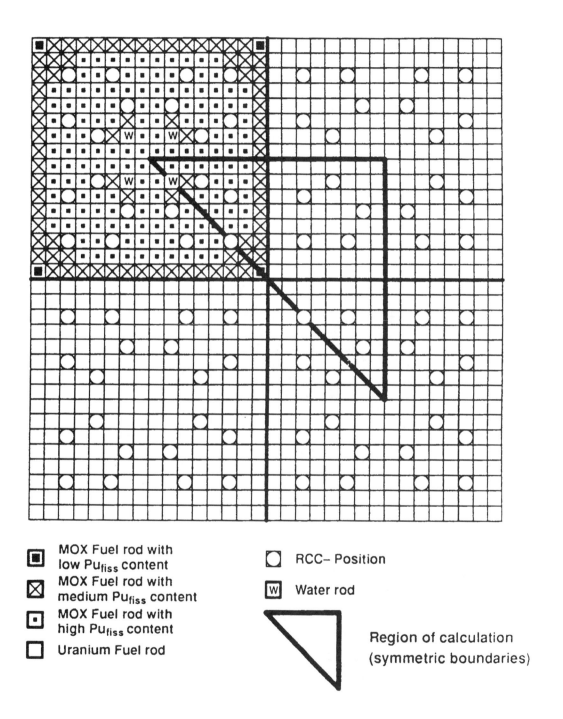

MOX Fuel rod with low Pu_{fiss} content

MOX Fuel rod with medium Pu_{fiss} content

MOX Fuel rod with high Pu_{fiss} content

Uranium Fuel rod

RCC– Position

W Water rod

Region of calculation (symmetric boundaries)

Figure 2.9 Macro-cell scheme of a MOX fuel assembly surrounded by UO_2 fuel assemblies

51

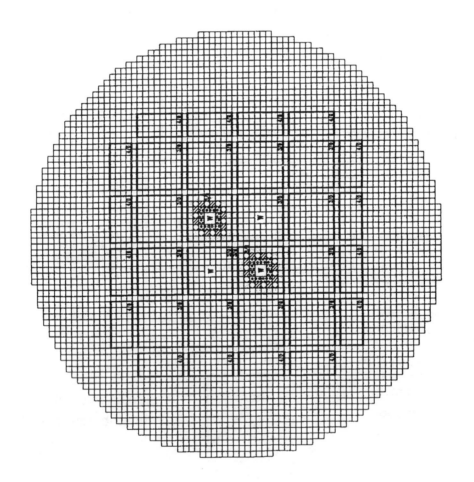

Figure 2.10 *VIP program for BWRs – schematic view and typical results*

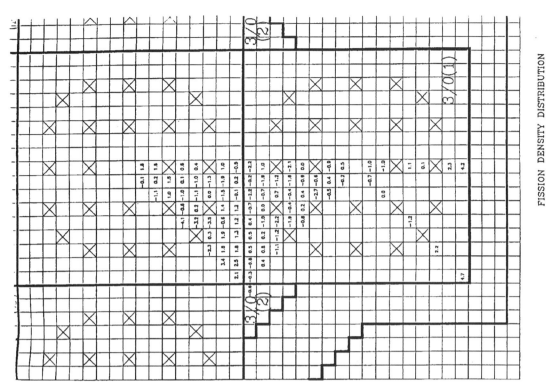

FISSION DENSITY DISTRIBUTION

Comparison Theory – Experiment (C–E)/E%

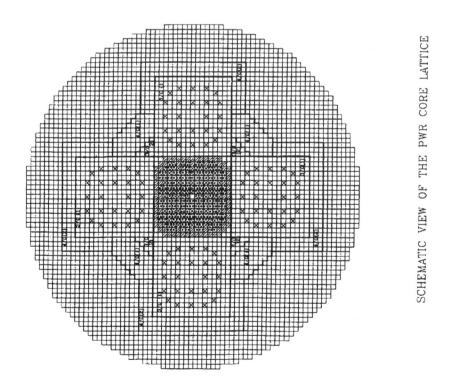

SCHEMATIC VIEW OF THE PWR CORE LATTICE

Figure 2.11 **VIP program for PWRs – schematic view and typical results**

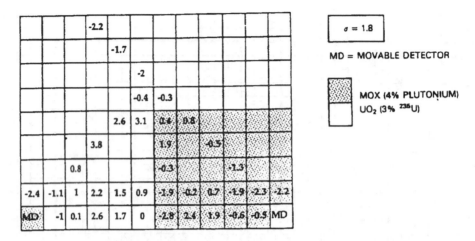

MINERVE:MELODIE B core: comparison of calculated and measured power.

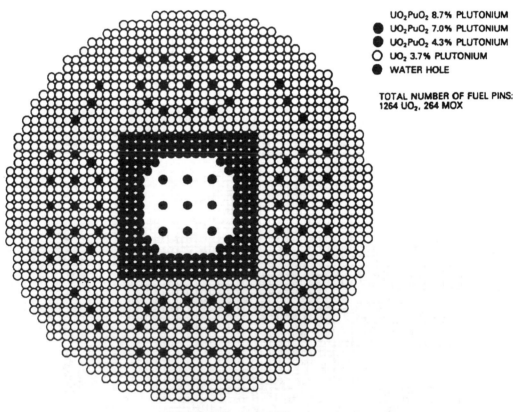

EPICURE: three-zone MOX assembly in a UO₂ core.

Figure 2.12 **Examples of MINERVE and EPICURE measurements**

54

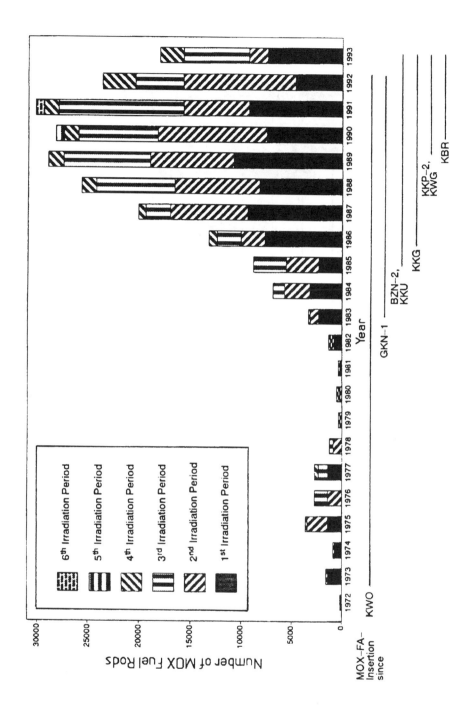

Figure 2.13 MOX insertion in PWRs by Siemens KWU Group until December 1993

55

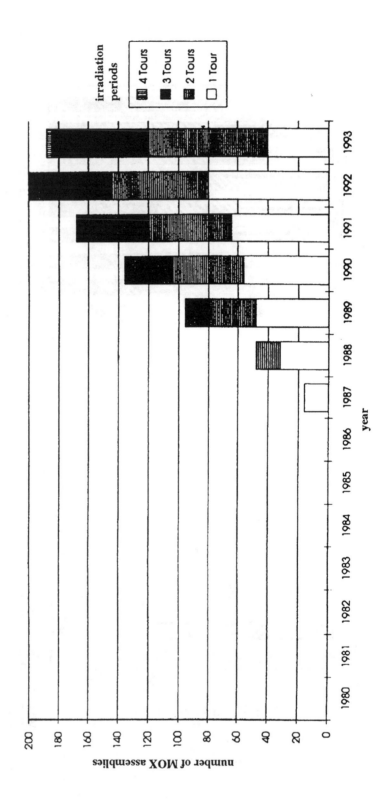

Figure 2.14 Plutonium recycling in a REP 900 EDF (PWR)

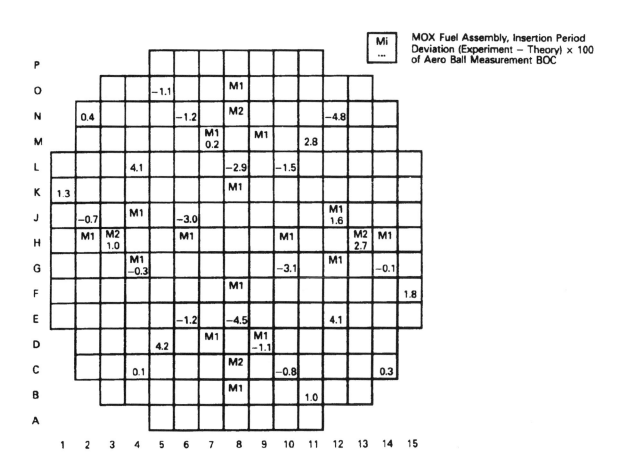

Mox loading pattern at KWG cycle 5 (20 MOX fuel assemblies): validation of power distribution at BOC

Figure 2.15 *Validation of power distribution for MOX-loading at Grohnde (Cycle 5)*

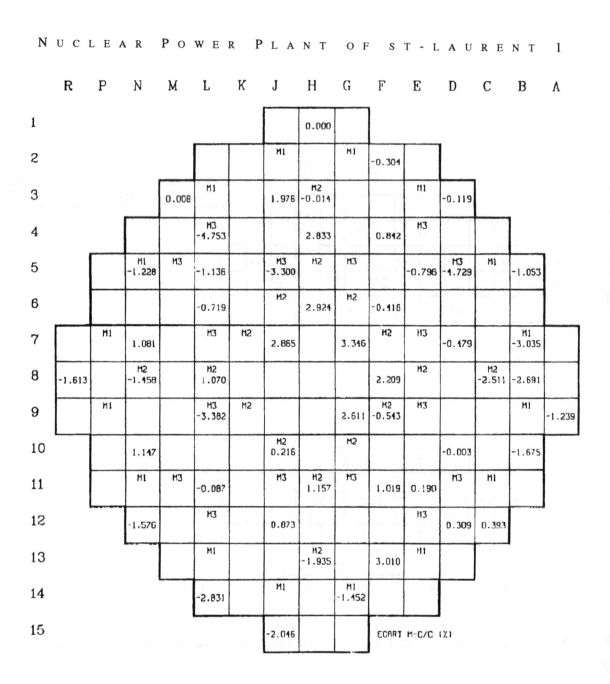

Average deviation (%) between measured and calculated activities (M-C)/C

M1 = MOX 1st irradiation – M2 = MOX 2nd irradiation – M3 = MOX 3rd irradiation

*Figure 2.16 **Validation of power distribution for St-Laurent 1, loaded with 3x16 MOX-FAs***

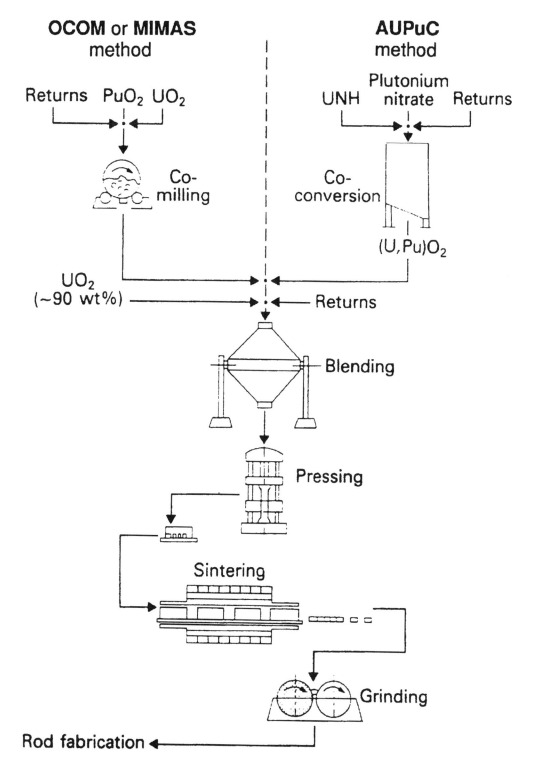

Figure 2.17 Flowchart for MOX fabrication for LWRs

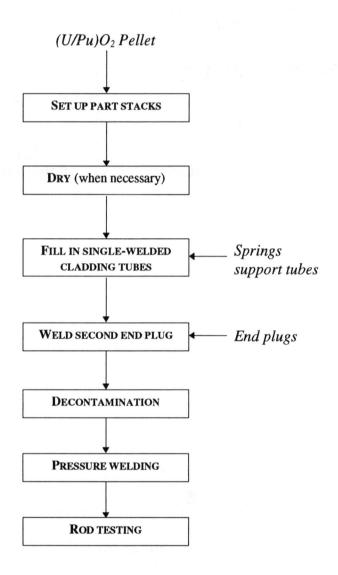

Figure 2.18 Flowchart for MOX fuel rod fabrication

U-235: $\nu_d = 0.01668 \pm 0.00070$ n/f		
Group	T½ (sec)	Relative Yield
1	54.51	0.038 ± 0.004
2	21.84	0.213 ± 0.007
3	6.00	0.188 ± 0.024
4	2.23	0.407 ± 0.010
5	0.496	0.128 ± 0.012
6	0.179	0.026 ± 0.004

U-238: $\nu_d = 0.0460 \pm 0.0025$ n/f		
Group	T½ (sec)	Relative Yield
1	52.38	0.013
2	21.58	0.137
3	5.00	0.162
4	1.93	0.388
5	0.493	0.225
6	0.172	0.075

Pu-239: $\nu_d = 0.00645 \pm 0.00040$ n/f		
Group	T½ (sec)	Relative Yield
1	53.75	0.038
2	22.29	0.280
3	5.19	0.216
4	2.09	0.328
5	0.549	0.103
6	0.216	0.035

Pu-240: $\nu_d = 0.0090 \pm 0.0009$ n/f		
Group	T½ (sec)	Relative Yield
1	53.56	0.028
2	22.14	0.273
3	5.14	0.192
4	2.08	0.350
5	0.511	0.128
6	0.172	0.029

Pu-241: $\nu_d = 0.0157 \pm 0.0015$ n/f		
Group	T½ (sec)	Relative Yield
1	54.0	0.010
2	23.2	0.229
3	5.6	0.173
4	1.97	0.390
5	0.43	0.182
6	0.2	0.016

Isotope	Delayed neutron fractions β
U-235	0.0067
U-238	0.0164
Pu-239	0.0022
Pu-240	0.0029
Pu-241	0.0054
Pu-242	0.0051

Table 2.1 Delayed neutron yield and half-life data [8] [9]

Categories of requirements ⇒	NORMAL OPERATION		ACCIDENTS
Areas of analysis ⇓	Reactor core	Spent fuel pool and new fuel store	Transients, LOCA, external events
Neutron Physics	MOX-FA-Design Core Characteristics	Sub-Criticality Decay Heat	Boron Worth Reactivity Coefficients Control Rod Worth
Thermal Hydraulics	unchanged	—	—
Systems Dynamics	Control Rod Worth	—	as above
Fuel Rod Design	Fission Gas Pressure Corrosion	—	Fuel Rod Failure Limit
FA Structure Design	unchanged	—	unchanged
LOCA Analysis	—	—	evaluated
Radiological Aspects	Activity Inventory	Activity Inventory Release Rates	Activity Releases

*Table 2.2 **Safety evaluations related to MOX fuel assembly licenses***

						Effect compared with uranium cores
MOX fuel assembly loading (number /%)	48/25	81/42	81/42	97/50	193/100	
Loading scheme	out-in-in	low leakage with gadolinium	low leakage with gadolinium	part low leakage	part low leakage	
Reload MOX and uranium fuel assemblies	16/48	24/32	24/32	24/24	64/0	
MOX fuel assembly type	16x16	16x16	16x16	18x18	18x18	
Fissile Pu content (wt%)	1.9/2.3/3.3	1.9/2.3/3.3	2.2/3.0/4.6	2.0/2.6/3.9/5.0	4.1	
U-235 content (wt%)	0.7	0.7	0.25	0.7	0.7	
Uranium fuel assembly U-235 enrichment (wt%)	3.4	3.5	3.5	4.0	–	
Cycle length (days)	329	310	318	323	454	same
MOX fuel assembly burnup						
. MOX batch (MWd/kg)	37.4	35.3	37.3	48.2	49.8	burnup
. maximum MOX fuel assembly (MWd/kg)	39.0	41.9	43.7	54.6	57.6	about the same
Initial boron concentration (ppm)	1247	1088	1085	1256	1996	lower
Reciprocal boron worth, BOC (ppm/% Δρ)	-135	-147	-158	-178	-298	higher ~ -120
MTC at EOC (pcm/K)	-59.5	-69.1	-61.4	-77.4	-78.5	higher ~ -55 to -65
Net control rod worth at EOC (% Δρ)	5.5	6.6	4.7	5.4	5.3	lower to same

*Table 2.3 **Equilibrium fuel cycles for large PWRs with MOX***

MOX fuel assembly loading (number /%)	264/31	0/0
Loading scheme	low leakage	low leakage
Reload		
MOX fuel assembly	40	0
average fissile plutonium content (wt%)	3.26	–
average U-235 content (wt%)	0.80	–
Uranium fuel assembly	96	136
average U-235 enrichment (wt%)	3.4	3.4
Cycle length, including coastdown (days)	296	298
MOX fuel assembly burnup		
MOX batch (MWd/kg)	45.2	–
maximum MOX fuel assembly (MWd/kg)	47.3	–
BOC hot excess reactivity (% $\Delta\rho$)	1.3	1.1
BOC cold shutdown margin (% $\Delta\rho$)	1.4	1.3
MCPR		
uranium fuel assembly	1.35	1.38
MOX fuel assembly	1.50	–
Maximum linear heat rate		
uranium fuel assembly (W/cm)	412	437
MOX fuel assembly (W/cm)	399	–

*Table 2.4 **Equilibrium fuel cycle for a large BWR with and without MOX***

EXPERIMENT					
NAME	FUEL	LATTICE TYPE	PITCH/INCH	RADIAL SAVING/CM	BUCKLING IN M^{-2}
Batelle [1] Critical	2 w/o PuO$_2$ inUO$_2$ 7.65 w/o Pu-240	triangular	0.85	–	–
"	2 w/o PuO$_2$ inUO$_2$ 23.50 w/o pu-240	triangular	0.85	–	–
Westinghouse [2] Critical	2 w/o PuO$_2$ inUO$_2$ 7.65 w/o Pu-240	quadratic	0.69	8.37	69.6 ± 1.0
"	"	"	0.69	8.37	68.7 ± 0.8
"	"	"	0.75	7.46 ± 0.09	90.0 ± 0.9
"	"	"	0.9758	6.95 ± 0.11	104.72 ± 0.86
"	"	"	0.9758	6.50 ± 0.15	107.11 ± 1.2
"	"	"	1.0607	6.76 ± 0.17	98.4 ± 1.2
"	"	"	1.38	6.46 ± 0.13	50.3 ± 0.3
"	UO$_2$ with 2.72 w/o U-235	"	0.69	–	–
"	"	"	0.9758	–	–

*Table 2.5 **Batelle and Westinghouse criticals***

---

[1] UNC-5211, [22].

[2] WCAP-3726-1, [23].

TYPE	CLAD	U-235 ENRICHMENT	PU CONTENT	PU ISOTOPIC COMPOSITION	NUMBER OF RODS
4/0	AISI-304	4.1 or 4.04%	0		1796
3/1.	Incoloy-800	3.00%	1.08%	B	425
2/2.7	AISI-304	2.00%	2.70%	A	476
2/3.1	AISI-304	2.00%	3.08%	A	32
1.76/3.2	AISI-304	1.79%	3.17%	A	65
0.7/5	AISI-304	U nat.	5.04%	A	63
0.7/4	AISI-304	U nat.	4.37%	C	26
3.3/0	Zircaloy	3.30%	0		1230
0.4/14.4	Zircaloy	0.35%	14.30%	F	230
0.3/9.7	Zircaloy	0.28%	9.70%	E	100
0.3/5.4	Zircaloy	0.25%	5.40%	D	66
0.6/4.8	Zircaloy	0.58%	4.80%	D	75
3.5/0 - 7.2	Zircaloy	0.035 [3]	0		60

Table 2.6 VENUS fuel inventory

Notes:

U-235 enrich =   U-235 / (U-235 + U-238) wt%

Pu content = PuO_2 / (UO_2 + PuO_2) wt%

Pu isotopic composition =   (Pu-238 / Pu-239 / Pu-240 / Pu-241 / Pu-242)

A	0.08/80.65/17.41/1.42/0.45
B	0.01/92.43/7.29/0.24/0.03
C	0.0/95.72/4.10/0.17/0.01
D	1.23/63.22/23.67/7.68/4.20
E	1.34/61.10/24.12/8.67/4.77
F	1.27/61.88/23.50/8.95/4.40

Decay of Pu-241 into Am-241 is not accounted for the indicated Pu isotopic composition

---

[3]   rods containing 7.2% Gd_2O_3 (Gd_2O_3 / UO_2 + Gd_2O_3).

PROGRAMME	TEMPERATURE IN °C	DESCRIPTION
KRITZ-1	20 - 210	a uniform U rod core
KRITZ-2	20, 245	pin cell cores with one or two uniform zones of U or Pu rods
	20, 230	BWR cores containing Pu island assemblies
KRITZ-3	20, 230	PWR cores with U or U and Pu with or without control rods
KRITZ-4	20, 245	BWR cores with U and BA (= Burnable Absorber)
	20, 245	BWR cores with U and BA with or without void in the central assembly

*Table 2.7 **Survey on KRITZ experiments at Studsvik /Sweden***

CORE	TEMPERATURE IN °C	WATER HOLES	FUEL IN CNETRAL ASSEMBLY	CONTROL RODES IN CENTRAL ASSEMBLY
1. U-WH1	22.0 229.0	small	UO_2	none
2. U-CR1	19.9 243.4	small	UO_2	AgCdIn rods
3. Pu-WH1	27.2 223.2	small	UO_2PuO_2	none
4. Pu-CR1	24.5 225.7	small	UO_2PuO_2	AgCdIn rods
5. U-WH2	27.4 228.7	large	UO_2	none
6. U-CR2	27.3 220.8	large	UO_2	thick B_4C rod

*Table 2.8 **Information about the cores in KRITZ-3***

Reactor (Type)	el. Power MW$_e$	Year
		'59 '60 '61 '62 '63 '64 '65 '66 '67 '68 '69 '70 '71 '72 '73 '74 '75 '76 '77
PRTR at PNWL (HW-PTR)	20	
Saxton (PWR)	8	
San Onofre-1 (PWR)	430	
BR-2 (PWR)	12	
BR-3 (PWR)	12	
SENA (PWR)	266	
MIHAMA 1 (PWR)	320	
Trino (PWR	260	
MZFR (HW-PWR)	57	
KWO (PWR)	345	
Big Rock Point (BWR)	72	
Quad Cities I (BWR)	809	
Dresden I (BWR)	200	
Garigliano (BWR)	160	
VAK (BWR)	16	
KWL (BWR) (Th Pu MOX)	252	
KRB-A (BWR)	252	

Table 2.9 Early MOX insertion programmes in LWRs

PARAMETER	CONFIGURATION	PLANT-CYCLE – FIRST TRANSITION			PLANT-CYCLE – SECOND TRANSITION	AVERAGE
		St Laurent B1, cycle 5	St Laurent B2, cycle 6	Gravelines 3, cycle 8	St Laurent B1, cycle 6	
critical boron concentration (ppm)	all rods out	-27	0	-5	+32	0
	R in	-20	+1	-11	+21	-2
	GG [4] in	-14	-14	-6	+24	-2
isothermal coefficient [pcm (°C)]	all rods out	+0.7	-0.8	+0.2	-0.4	-0.1
	R in	-0.3	-0.9	+0.1	-1.0	-0.5
	GG in	+1.5	-0.6	-0.5	-3.3	-0.7
control bank worth (%)	R	+2.8	-2.5	+3.4	+2.3	+1.5
	G1	-5.2	+2.5	+11.2	+1.3	+2.4
	G2	+0.1	+3.2	+2.1	+1.5	+1.7
	N1	+4.2	-0.6	+2.9	+4.7	+2.8
	N2	+7.7	+0.6	+4.3	+3.4	+5.3
	SB	+5.1	+4.1	+5.9	+0.8	+4.0
	SC	-4.3	-1.1	+4.2	+3.8	+0.6
	SA + SD2	+1.9	-0.7	+1.3	+7.0	+2.4
	SD1(SA+SD2in)	-6.4	+4.5	+7.7	+3.1	+5.4
	N2-1(SA+SD2in)	+7.9	+4.9	+6.8	+3.1	+5.7
	GG in	+4.1	-6.4	+4.1	-0.2	-0.1

*Table 2.10 **Validation of MOX-insertion in French PWRs***

---

[4] GG = gray control banks (G1 + G2 + N1) at zero-power position.

Chapter 3

ANALYSIS OF PHYSICS BENCHMARKS FOR PLUTONIUM RECYCLING IN PWRS

3.1 Introduction

The recycling of plutonium in PWRs in MOX fuel assemblies is a technology which is now well established and many countries have many years' experience to draw on. As discussed in Chapter 2, it is fair to say that within the constraints of current fuel management schemes, discharge burnups and plutonium isotopic vectors, physics methods are available, which can be considered to be mature and fully proven.

The validity of present methods cannot be assumed to extend outside the current constraints, however, and further validation will be required to demonstrate that both the basic nuclear data and the calculational methods remain adequate for the more challenging problems that are expected to arise within the next decade. The challenges to existing physics methods will come from high burnup fuel management schemes and feed plutonium with lower fractions of the fissile isotopes Pu-239 and Pu-241. The effect of both these changes will be to increase the total plutonium loading necessary in the MOX fuel. This will increase thermal neutron absorption and drastically alter the thermal neutron spectrum.

Unfortunately, experimental validation will not be forthcoming for the new situation for several years, yet it is important to have some indication of what level of development effort will be required to address the possible shortcomings of present physics methods. Faced with this situation, the WPPR agreed that a set of benchmark exercises would be a valuable means of making progress in the interim period before any practical results become available from in-reactor irradiation experience. As mentioned in Chapter 2, a set of three benchmarks were devised and solutions submitted from a large number of contributors. It was hoped that a comparison of the results would give valuable insights into the likely requirements as regards improving the nuclear data and methods. While accepting that such benchmarks could not possibly identify the 'true' answer, it was anticipated that a consensus view on the most probable answers would emerge which would be helpful in guiding future work.

This chapter presents a detailed analysis and critique of the results of the three PWR benchmarks and presents the principal results in a convenient summary form. Because of the large number of participants and the complexity of the physics solutions, it is not practical to present the full set of results. These have, however, been compiled in separate volumes, [1] [2], for the benefit of those with a detailed working interest, which are available on request from the OECD/NEA. These also include working papers that will be of value to specialists interested in analysing the results in detail.

3.2 Objectives of PWR MOX benchmarks

Three benchmarks were devised for MOX in PWRs. The first two are simple infinite array pin cell problems designed to allow intercomparison of infinite multiplication factors as a function of burnup. The first such pin cell problem, designated 'Benchmark A' comprises a pin cell with plutonium of a low isotopic quality (i.e. a low fraction of the thermally fissile isotopes Pu-239 and Pu-241). It is expected that such plutonium will become available for recycle at some future date, when MOX fuel assemblies are themselves reprocessed. The quality of plutonium recovered from PWR spent fuel decreases during each recycle, the rate depending on the discharge burnup of the reactor fuel cycle and on the ratio in which MOX assemblies are co-processed with UO_2 assemblies in the reprocessing plant.

The particular isotopic composition specified for Benchmark A represents a hypothetical case of the fifth recycle of plutonium for a scenario in which the MOX assemblies are blended with UO_2 assemblies in a ratio which reflects that which will arise in a self-generation recycle mode in a PWR. The total plutonium content is 12.5 w/o (6.0 w/o fissile) and the isotopic vector is as follows:

Pu-238	Pu-239	Pu-240	Pu-241	Pu-242
4%	36%	28%	12%	20%

The poor plutonium isotopic quality in Benchmark A demands a high concentration of total plutonium in order to compensate for neutron absorption in Pu-240 and Pu-242 isotopes. The high plutonium concentration poses a severe challenge to existing nuclear data libraries and lattice codes, which was the driving force behind the specification.

The other pin cell problem, designated 'Benchmark B' specified a plutonium isotopic vector with a higher fissile fraction that is representative of commercial PWR MOX recycle at the present time. The total plutonium content is 4.0 w/o (2.8 w/o fissile) with the following isotopic vector:

Pu-238	Pu-239	Pu-240	Pu-241	Pu-242
1.8%	59%	23%	12.2%	4.0%

This problem was intended to act in the form of a 'control' to show whether the spread of results in the more challenging problem could be attributed to the poor quality plutonium vector or to underlying differences in the nuclear data and methods which also apply to today's situation.

The final problem, designated the 'Void Reactivity Effect Benchmark', specifies a more complicated geometry corresponding to a supercell configuration of a 30x30 array of PWR fuel cells, with reflective boundary conditions. The central 10x10 region (see Figure 3.1) consists of either UO_2 or MOX rods (with three different plutonium contents), the configuration of which alternates between full moderator density and complete voidage of the moderator. In every case the outer part of the macrocell, which consists entirely of UO_2 pincells, is assumed to be fully moderated. This supercell is an idealisation of a series of experiments that were recently carried out in the VENUS experimental reactor at Mol, Belgium, as part of the VIP-O international collaborative programme. The intention was to specify a problem in which it would be possible to compare the void reactivity defect calculated by various nuclear data libraries and codes.

The void coefficient is a very important parameter in any water reactor. It determines the reactivity feedback associated with steam formation, which for inherent safety should always be negative (i.e. an increase in void decreases the core reactivity). At the high plutonium concentrations expected with poor plutonium isotopic quality, it was known that the void coefficient in mixed UO_2/MOX lattices becomes much smaller in absolute magnitude and at some point would become positive. Calculating the point at which the void coefficient changes sign is a difficult problem, as there is a delicate balance between the opposing effects of moderation and absorption and the plutonium resonances also play a significant complicating role. Ensuring that the void coefficient inherent in the MOX regions of a PWR core remains negative will be an important constraint determining the limits to multiple recycle in PWRs, as it may prove unacceptable for licensing to have regions in the core where the inherent void reactivity feedback is positive.

The void coefficient issue first arose in connection with studies of tight lattice PWRs which were carried out some 10 years ago. It does not normally cause any concern for standard lattice PWRs, with the modest plutonium concentrations required in the context of single recycle. With multiple recycle, however, the poor plutonium isotopic makeup demands high plutonium contents and assuring a negative void coefficient becomes an important factor.

For a full specification of the void reactivity effect benchmark, refer to Volume 3, [2]. Section 3.4 provides a detailed description and discussion of the void reactivity effect benchmark.

3.3 Infinite lattice benchmarks

A complete specification of the two infinite lattice benchmark problems, designated Benchmarks A and B, can be found in Volume 2, [1].

3.3.1 Participants, methods and data

A total of 14 solutions were contributed for Benchmark A and 13 for Benchmark B, representing 12 institutions from 9 countries. A full list of all the contributors is provided in Table 3.1. It identifies the codes and nuclear data libraries used by the various contributors and where necessary makes pertinent remarks. The acronyms for the institutes or organisations will be used to identify each contributor throughout the chapter.

The two CASMO-3 solutions were withdrawn as the cross-section libraries used were not suited for high plutonium fuel as specified in the benchmark.

As can be seen most conveniently from Table 3.1, most of the contributors to Benchmarks A and B used deterministic lattice codes. These are the usual tools used for nuclear design applications such as calculating reactivities and irradiation depletion effects. Some contributors used Monte Carlo methods, which provide a useful cross-check on the methods, but which cannot carry out depletion calculations and are therefore restricted to the zero burnup step. The Monte Carlo codes are also restricted in that nuclear data tabulations are only usually available for a limited set of materials temperatures. Table 3.1 highlights where the temperatures available did not coincide with the benchmark specifications.

INSTITUTE	COUNTRY	CODE	DATA BASE/LIBRARY	No OF GROUPS	REMARKS
ANL	U.S.A.	VIM	ENDF/B-V	infinite	300 K, Zircaloy, no depletion
BEN	Belgium	LWRWIMS	1986 WIMS	69	
BNFL	U. K.	LWRWIMS	1986 WIMS	69	
CEA	France	APOLLO-2	JEF-2.2 CEA 93	172	
ECN	Netherland	WIMS-D	JEF-2.2 SCALE	172	
EDF	France	APOLLO-1	CEA 86	99	
Hitachi	Japan	VMONT	JENDL-2/ENDF/B-IV	190	
IKE-1	Germany	CGM/RSYST	JEF-1	224/45	
IKE-2	Germany	MCNP 4.2	JEF-2.2	infinite	300 K/600 K, no depletion
JAERI	Japan	SRAC	JENDL-3.1	107	
PSI-1	Switzerland	BOXER	JEF-1	70	
PSI-2	Switzerland	CASMO-3	ENDF/B-IV	40	Zircaloy, withdrawn
Siemens	Germany	CASMO-3	J70	70	withdrawn
Studsvik	Sweden	CASMO-4	JEF-2.2	70	

Table 3.1 **Summary of participants in pin cell benchmarks**

At this stage it is appropriate to draw attention to some of the special physics aspects that need to be accounted for in the MOX benchmark calculations and to highlight aspects which participants took particular care to model rigorously:

The relatively large thermal absorption cross-sections of plutonium reduces the thermal neutron flux considerably compared with uranium, while the flux at higher energies is less drastically affected. The result is that the neutron spectrum in a MOX assembly is much harder than that in a UO_2 assembly and the resolved resonances have a much higher impact on the calculation of group cross-sections. In addition, the unresolved resonances and the threshold reactions in the MeV range also require more careful attention. Some of the contributors used codes where resonance self-shielding in all plutonium isotopes is treated rigorously, and this has an important bearing on the results, as will be seen later.

The energy per fission values to be used were defined in the benchmark specifications, with five isotopes only contributing to energy release. Only six of the participants used the specified values and only the EDF and CEA solutions omitted the energy release from other isotopes. The other participants calculated the energy production according to their normal design methods, which account for all fissile contributions. The effect is that the EDF and CEA solutions have a slightly stretched effective burnup scales.

Most participants took account of (n,2n) reactions by lowering the absorption cross-sections artificially. The effect increases the multiplication factor by about 0.2%. A rigorous treatment, however, involves modifying the actinide chains explicitly and shows that artificially reducing the absorption cross-sections introduces for higher burnups a systematic error due to the higher levels of Np-237 which build up. The influence of the fission product spectrum is of the same order.

3.3.2 Results

Figures 3.2.A and 3.2.B show compilations of k-infinities for Benchmarks A and B respectively from the various participants. Tables 3.2.A and 3.2.B list the same data. These are the principal results of the benchmarks. The spread of results at zero irradiation is 3.1% for Benchmark A and 1.3% for Benchmark B. There is also some spread in the slope of k-infinities versus burnup. The corresponding values for 50 MWd/kg are 4.9% and 2.9% respectively.

Detailed comparisons (both tabular and graphical) of the burnup dependence of the actinides and fission products number densities, absorption rates, fission rates, neutrons per fission and neutron spectra are displayed in Volume 2, [1].

3.3.3 Discussion

Multiplication factors

Referring to Figure 3.2.A, it can be seen that there is a disappointingly large spread of k-infinities for Benchmark A (approaching 3.1% at zero burnup), although it is encouraging that there is substantial agreement as to the slope of k-infinity with burnup. Some of this spread is, however, straightforward to account for.

Not all current lattice codes are able to represent resonance absorptions in the higher plutonium isotopes. This is because historically the absolute concentrations of the higher plutonium isotopes in both UO_2 and MOX fuels have always been low enough that self-shielding in them was less important. As explained earlier, the purpose of Benchmark A was to test code predictions in a challenging situation where this no longer applies. Thus Benchmark A specifies 3 w/o *absolute* of Pu-242, for which self-shielding can by no means be neglected. In view of this, it is not surprising that some of the results are systematically in error. For the conditions of Benchmark A, the effect is estimated to be worth a systematic bias of about 2.5% in k-infinity, so that the code predictions in which higher isotope self-shielding is not applied should be *increased* by this amount. The solutions provided by BEN and BNFL (both LWRWIMS) fall into this category. From Figure 3.2.A, it is apparent that if these contributions are corrected upwards by 2.5%, or if only those codes with rigorous higher isotope self-shielding are included, the spread of results is considerably narrowed to about 0.9-1.5% depending on the burnup.

Considering the solutions incorporating rigorous self-shielding, the 0.9% spread in k-infinities most probably arises from underlying differences in the nuclear data libraries. Other physical effects could contribute to this spread, such as the variation of shielding factors within the fuel pin, or the variation of those factors with burnup. These effects have been calculated with APOLLO-2, the results of which can be found in Volume 2, [1]. There is a clear tendency for solutions based on a common data library to be very close, for example PSI-1 and IKE-1 (both JEF-1) as one sub-group , CEA, ECN, IKE-2 and STU (all JEF-2) as a second sub-group and HIT and JAE (both JENDL) as the third. This suggests that differences in the lattice code methods are less important than the nuclear data evaluations.

In respect of the 1.5% residual spread, it has to be said that if this was representative of the uncertainty on the lattice calculations, it would be unacceptable for design and licensing applications. Current nuclear design methods typically claim uncertainties on reactivity of about 0.2%, with occasional outliers of up to 0.5%. A concerted effort will clearly be necessary to resolve the outstanding differences and this will necessitate experimental validation. The situation is particularly unsatisfactory because the reactivity of MOX fuel tends to increase only very slightly as the plutonium content increases, an effect which is greatly exaggerated in the Benchmark A situation because of the low fissile fraction of plutonium. Thus, any attempt to increase reactivity by loading a higher fissile plutonium content is to a large extent opposed by the increased absorption from the even isotopes. This means that any uncertainty in the reactivity predictions will translate into a disproportionately large spread in the plutonium concentration needed to achieve the specified lifetime reactivity.

The codes and libraries give better agreement for the more conventional MOX fuel specified for Benchmark B (see Figure 3.2.B). The same grouping of solutions as in benchmark A is also visible in Table 3.2.B.

The IKE-2 solution was carried out with a fuel temperature 33.2°C lower than specified, a clad temperature 20.6°C higher than specified and a moderator temperature 5.9°C lower. Sensitivity calculations with APOLLO-2 indicate that the combined effect of these temperature differences is that the MCNP k-infinity is too high by 93 and 85 pcm for Benchmarks A and B respectively, so that the corrected k-infinities are 1.1298 for Benchmark A and 1.1840 for Benchmark B. The same problem arose with the ANL solution, which was carried at room temperature; the figures quoted in Volume 2, [1] with the detailed results were corrected using the IKE-2 results.

The problem of different models of energy release mentioned in the previous section affects the burnup scale because of the omission of the contributions of fissionable isotopes, mainly Pu-238 and Pu-240. The effect is nearly independent of burnup. The stretching factor for the results of CEA and EDF is about 1.03 and 1.01 for Benchmark A and B, respectively. Sensitivity calculations of CEA (see Volume 2, [1]) gave correction values of -392 pcm and -196 pcm for benchmark A and B respectively in order to make the results of CEA and EDF at 50 MWd/kg comparable with the others.

Reactivity change with burnup

The reactivity change with burnup in Benchmark A is moderately consistent between the various contributions with a spread of 2.5% Δk at the highest burnup step. When only the results from the codes which are more established in terms of commercial MOX experience are included, the spread reduces to about 1.5% Δk. There is a tendency for those contributions in which Pu-242 self-shielding was not modelled to have the highest reactivity swings (e.g., BEN and BNFL). This may be attributable to the higher levels of Am-243 which this implies, since Am-243 has higher absorption cross-section than Pu-242. This is borne out by the absorption rate plots (see Volume 2, [1]), which shows that both the BEN and the BNFL solutions have the highest absorption rates in Am-243.

For Benchmark B the spread of Δk, at 2.2%, is only slightly smaller than that of Benchmark A. This implies that the bulk of the discrepancy arises from inherent differences in the depletion characteristics, probably deriving from nuclear library differences and differences in the resonance calculational models.

One-group fluxes

The one-group fluxes also show discrepancies, spreads of approximately 7% and 4% applying to Benchmarks A and B respectively. This may stem in part from the fact that not all contributors were able to use the specified MeV/fission values, because such a facility is not normally provided in lattice codes. It is surprising that differences exist even for those contributions in which the specified MeV/fission values were used.

Absorption rates

The normalisation of the flux in the cell according to the usual condition 'total absorption in cell equal to unity' ensures that the error in the absorption rate is equivalent to the error in k-infinity, but with the opposite sign. Consequently, the macroscopic absorption rates of individual isotopes in the fuel can be used to correlate the differences in k-infinity to individual isotopes. Actinides with a significant configuration to the absorption rates (> 1%) having spreads worth noting at zero irradiation or at 50 MWd/kg are: U-235, U-238, Pu-238, Pu-239, Pu-240, Pu-241, Pu-242, Am-241, Am-243 and Cm-244. The largest discrepancies are for Pu-242, consistent with inappropriate treatment of the 2.7 eV resonance in some of the solutions, in which the bulk of the Pu-242 absorptions occur. Naturally, Benchmark A shows by far the largest discrepancy, due to the high absolute concentration of Pu-242.

Relatively large spreads are also noticeable for U-238 in both benchmarks. Since the U-238 cross-sections are well known, it is likely that the resonance absorption calculational methods are responsible.

The mean absorption rates for the principal fission products also show spreads. The absorption rates are for the most part lower than 1%, but the spread of values are often nearly as large as the rates themselves. There is the potential for these spreads to contribute to an uncertainty of up to 1% in k-infinity, and this may arise from a combination of uncertainties in the nuclear cross-sections, fission yields and depletion models.

Fission rates and neutrons per fission (ν)

The variations in fission rate cause differences in k-infinity, the largest being for Pu-239, Pu-241 and U-238. The reason for the differences seen in U-238 may be due to both the use of fission spectra which are inappropriate for the actual composition and the inadequate cross-section data.

For both benchmarks large variations in ν for the minor actinides and differences in the percent range are seen for the main actinides. The spreads on U-238, Pu-239 and Pu-241 are sufficient to cause uncertainties of the order of 0.1% Δk in k-infinity.

Number densities

The discrepancies in number densities are in most cases higher than those seen in reaction rates. The principal actinides all fall within 10%, except for Pu-242, for which it is about 20%. The concentrations of Am-243 and the Cm isotopes show similar deviations. The spread for Am-243 must partly be due to the self-shielding issue, as discussed earlier. The minor actinides also show large variations. Overall, the situation is not acceptable, especially for Benchmark A.

3.3.4 Conclusions – infinite cell benchmarks A & B

The set of resulting multiplication constants as a function of burnup shows large fluctuations of up to 4.9% and 2.9% for the pin cells of benchmark A and B, respectively. The solutions with higher dispersion are calculated by commercially established codes, which are mainly applied and verified for uranium fuel. If these solutions are not included, the spread decreases to 0.9 % at BOL and 1.5 % at 50 MWd/kg. Most participants, whose results represent this narrower band, applied new data bases and refined resonance calculations for the generation of shielded resonance cross-sections. The situation is similar to the one encountered at the OECD/NEA meetings on High Conversion LWR benchmarks investigating the behaviour of water moderated MOX fuel [3]. The main resulting recommendations of these meetings are valid for the present benchmark also and are the following [4]:

- The calculational methods have to take into account resonance shielding, and should include mutual shielding, over the whole energy region for the fuel and cladding nuclides and the major fission products;

- Basic nuclear data of sufficient quality are needed, in particular for U-238 and the Pu isotopes, but also for higher actinides and fission products.

One part of the spread in the results of the present benchmarks originates from differences in the applied data. Solutions, in which the new data bases JEF-2, JENDL-3 and ENDF/B-V are used, show characteristic discrepancies for instance in the specific reaction rates, which should be correlated not only to differences in cross-sections of specific isotopes but also in cross-sections in specific energy regions. The energy integrated reaction rates provided in the benchmark do not give sufficient information to make a detailed evaluation in this respect possible. Energy dependent reaction rates would help in identifying where data from evaluations need improvement and also in refining the methods for calculating weighting spectra and weighted cross-sections.

The large uncertainty related to the minor actinide production is worthwhile mentioning as a by-product of this benchmark. Its origin stems both from differences in the cross-section bases and from inadequate resonance shielding calculations (neglecting the mutual shielding effect). This appears clearly in the differences of the Pu-242 number densities and its successors Am-243 and Cm-244.

A supplementary benchmark, aimed at clarifying the applied cross-section processing methods for users of JEF-2 and JENDL-3 evaluated data libraries, was initiated at the meeting in November 1994. Details are reported in Volume 2, [1] [5].

3.4 Void reactivity effect benchmark

A complete specification of the void reactivity effect benchmark problem can be found in Volume 3, [2]. This benchmark specifies a supercell configuration of a 30x30 array of PWR fuel cells, with reflective boundary conditions. The central 10x10 region (see Figure 3.1) consists of either UO_2 or MOX rods (with three different plutonium contents) whose configuration alternates between full moderation and complete voidage of the moderator. In every case the outer part of the macrocell, consisting entirely of UO_2 cells, is assumed to be fully moderated. The configurations with UO_2 rods (3.35 w/o U-235), MOX rods of high enrichment (14.4 w/o total Pu), MOX rods of medium enrichment

(9.7 w/o total Pu) and MOX rods of low enrichment (5.4 w/o total Pu) in the central 10x10 sub-assembly are designated UO₂, H-MOX, M-MOX and L-MOX respectively.

3.4.1 Participants, methods and data

Eighteen solutions were contributed to the void reactivity effect benchmark from 12 institutions representing 8 countries, of which 5 contributions were based on Monte Carlo codes. A summary of the codes and nuclear data libraries used is provided in Table 3.3, together with the acronyms identifying the contributors.

INSTITUTE	COUNTRY	CODE	DATA BASE/LIBRARY	No OF GROUPS	REMARKS
ANL	U.S.A.	VIM	ENDF/B-V	infinite	Zircaloy
BEN	Belgium	LWRWIMS	1986 WIMS	69	
CEA-3	France	APOLLO-2	JEF-2.2 CEA 93	172/99	
CEA-4	France	ECCO-52/ERANOS	JEF-2.2 CEA 93	1968/172	
CEA-5	France	APOLLO-1	ENDF/BV + JEF-1 CEA 86	99	
CEN	Belgium	DTF4/DOT-3.5	MOL-BR2	40	
ECN	Netherland	MCNP-4.2	JEF-2.2 SCALE	172	
ENEA	Italy	MCNP-4.2	JEF-1	infinite	
Hitachi	Japan	VMONT	JENDL-2/ENDF/B-IV	190	
IKE-1	Germany	CGM/RSYST	JEF-1	224/60	
IKE-2	Germany	MCNP-4.2	JEF-2.2	infinite	
IPPE	Russia	WIMS/D4	FOND-2 WIMS/ABBN		withdrawn
JAERI-1	Japan	SRAC	JENDL-3.1	107	
JAERI-2	Japan	MVP	JENDL-3.1	infinite	
JAERI-3	Japan	SRAC/PIK	JENDL-3.1	107	
JAERI-4	Japan	SRAC/MOSRA	JENDL-3.1	107	
Siemens	Germany	CASMO-3	J70	70	withdrawn
Toshiba	Japan	MCNP-4.2	JENDL-3.1	infinite	

*Table 3.3 **Summary of participants in void reactivity effect benchmark***

The results obtained with CASMO-3 and WIMS/D4 were withdrawn.

3.4.2 Results

Tables 3.4 and 3.5 give k-infinities for the *central 10x10* sub-assembly considered in isolation as an infinite lattice, for the fully moderated and fully voided configurations respectively. These k-infinities are an useful indication of the underlying agreement of the nuclear data libraries and fine-group flux calculational methods, without complications arising from neutron leakage to and from the 10x10 sub-assembly. The bottom line in both of these tables gives the arithmetic averages of all the contributions. Figures 3.3 and 3.4 display the same information in graphical form, while Figure 3.5 shows the corresponding void defects, being the difference in k-infinity between the fully moderated and fully voided situations.

Referring to Table 3.4, the low k-infinities for the MOX configurations (when compared with the UO$_2$ configuration), reflect the increased thermal absorption in MOX assemblies. Comparing these with the corresponding voided k-infinities from Table 3.5, it can be seen that the k-infinity for the UO$_2$ lattice decreases considerably in the voided situation, corresponding to a negative void reactivity defect. While the same is true for the L-MOX and M-MOX cases, the negative defect is smaller in magnitude, the H-MOX case has a higher k-infinity in the voided case, corresponding to a positive void reactivity defect. The reason for this is not difficult to understand when it is considered that the H-MOX case, consisting of 14.4 w/o total plutonium with no moderator resembles a fast reactor more than a water reactor, so that it is no surprise that k-infinity has a high value. In the fast reactor-like spectrum, all the plutonium isotopes contribute to fissions and k-infinity increases almost linearly with total plutonium content. In contrast, in the fully moderated situation only the odd plutonium isotopes are fissionable and k-infinity increases much more slowly with increasing plutonium content due to the increasing contribution of absorption in the even isotopes.

Tables 3.6 and 3.7 list k-infinity values for the whole macrocell for the fully moderated and voided cases respectively. Figures 3.6 to 3.8 show the k-infinities and void defects in graphical form. Since the macrocell volume is largely composed of UO$_2$ pins and only one ninth of the pins are in the central sub-assembly, the overall k-infinity varies considerably less between the various configurations. A particular point to bear in mind is that only the central 10x10 sub-assembly is subject to voiding and that there is a significant source of thermal neutrons into the voided sub-assembly associated with the surrounding fully moderated UO$_2$ region.

Although the all-UO$_2$ macrocell shows a negative void reactivity defect, the averages for the three MOX cases all show positive void reactivities. This should not be taken to imply that the contribution of MOX assemblies to void reactivity will be positive in the situation of a real reactor, except with high plutonium contents. There is no possibility of the situation in a reactor arising where one assembly is fully voided while its neighbours are fully moderated. In the artificial situation modelled in this benchmark the role of thermal neutrons leaking from the fully moderated region into the voided region is crucial. In a fully moderated MOX assembly, the thermal neutron diffusion length is such that the boundary effect from leakage from the UO$_2$ to the MOX region extends over at most two or three rows of fuel pins. In the voided case the thermal neutron diffusion length increases so that the transient thermal neutron current extends throughout the 10x10 sub-assembly. The effect of voiding is therefore to couple the central MOX rods more strongly to the UO$_2$ regions which are rich in thermal neutrons. The central MOX rods are thereby able to contribute more effectively to the overall multiplication factor in the voided condition.

This result has been analysed extensively and a full account of several different approaches to understanding it, is included in Volume 3, [2].

Figure 3.9 shows a typical fission density traverse for one of the macrocell cases (a diagonal traverse for a macrocell with H-MOX rods in the central 10x10 region). The spread of values in the voided region is large, reflecting the severe interface effect between the fully moderated UO$_2$ region and the fully voided MOX region that is challenging to calculate.

3.4.3 Discussion

The multiplication factors of the infinite cell lattices show spreads of up to 4% (Tables 3.4 and 3.5 and Figures 3.3 and 3.4). If the outlying solutions are omitted the spread reduces to about 1% for the moderated cells and 2% for the unmoderated cells. The reason for the latter may in some cases be partly due to deficiencies in the spatial coupling methods or in approximations inherent with cylindricising the square cells. The fact that there is a significant spread for the moderated cells is indicative of differences in the nuclear data libraries. There is a clear tendency for solutions based on the same evaluated nuclear data set to be very close; as exemplified in the case of solutions based on JEF-2.2 in Volume 3, [2], the overall deviations are reduced to half. That the spread is due to the nuclear data sets is reinforced by the observation that even the Monte Carlo solutions have a significant spread; for the Monte Carlo codes there are no concerns over approximations in the pin-cell homogenisation or in the transport method, so that the spreads are directly attributable to the nuclear data. Thus, for example, the JEF-2 evaluated dataset yields larger k-infinities than the average for the UO$_2$ fuel and lower k-infinities for MOX, whereas JENDL-3 shows the opposite tendency.

It should be borne in mind, however, that in the voided situation the unresolved resonance self-shielding becomes more important because of the shift in the spectrum to higher energies. Neglecting the self-shielding in this energy range results in a k-infinity too low by about 1.5%. The main contributor is, as might be expected, U-238.

The multiplication factors for the whole macrocell, given in Tables 3.6 and 3.7, reflect the spreads seen on the infinite lattice results for uranium cells, since uranium rods make up a fraction 8/9 of the volume in the MOX configurations and these rods are fully moderated in all cases. The Monte Carlo methods are particularly valuable reference datapoints, as they allow an exact representation of the strong spatial dependence of the flux spectra in the vicinity of the voided/moderated boundary. Deterministic methods which rely on homogenising pin cells are particularly questionable at this boundary, especially if zero current is assumed at the pin cell boundaries in generating the homogenisation spectra.

The fission density curves, of which Figure 3.9 is a typical example, show satisfactory agreement for the fully moderated cases, but show apparently highly discrepant results for the voided cases as plotted. This is to some extent an artefact of normalising each curve at the centre pin, which although convenient for presenting the results, transfers the spreads entirely to the UO$_2$ region. In reality, the power distribution errors will be much smaller, as the spread really applies to the voided region. Since the fission rates are very small in the voided region, the absolute errors on fission rates there are within 10%. As might be expected the largest deviations actually occur at the MOX pin which borders on the UO$_2$ region. It is in this location where transport and homogenisation errors are most significant with the deterministic methods. The Monte Carlo codes again provide a useful reference, since they are not subject to such errors and the spread of results from the Monte Carlo codes is indicative of that in the underlying nuclear data.

The flux spectra of the central MOX pins show marked depressions in the voided cases at the resonance energies. The thermal spectra of the central pins are also very highly depressed, but those of the corner and edge pins are much closer to those of the adjacent UO$_2$ rods due to the neutron current from the latter [2].

3.4.4 Conclusions

The objective of the void reactivity effect benchmark was to compare the performance of the nuclear data libraries and codes presently available on a problem involving the calculation of the void reactivity effect in a mixed UO₂/MOX macrocell. The spread of k-infinities for the fully moderated infinite lattice amounts to somewhat greater than 1%, which is considered excessive and should be improved. Solutions obtained with the same nuclear data libraries tend to be grouped together, which indicates that the differences in the libraries are largely responsible. The voided configurations show larger spreads which are probably partly the result of using approximate homogenisation and transport methods at the voided/moderated boundary and partly due to differences in higher energy cross-sections, which assume a greater importance in the voided situation.

Overall, however, there is substantial agreement as to the trend for the void reactivity effect to become more positive as the plutonium content of the MOX region increases. The infinite lattice results show that the inherent void reactivity of MOX assemblies becomes positive somewhere between 10 and 14 w/o total plutonium content, at least with the isotopic composition assumed here.

Similar results were obtained with the hybrid UO₂/MOX macrocell, which are dominated by the properties of the moderated UO₂ pins which make up the bulk of the macrocell. The same trend towards more positive void reactivity effects as the plutonium content of the MOX regions increases was established. Because of the artificial nature of the problem, with a fully voided MOX sub-assembly adjacent to a fully moderated UO₂ driver region, positive void reactivity defects were obtained even for the L-MOX case, a result which would not occur in the more realistic case of more uniform void distribution.

Following a detailed examination of the results, it is recommended that the differences in the actinide cross-sections in the JEF-2.2 and JENDL-3.1 nuclear data libraries should be evaluated closely to explain the differences seen in the Monte Carlo calculations.

3.5 Benchmark analysis – Overall conclusions

When the WPPR proposed this set of benchmark exercises, it was expected that good agreement between the various solutions would be obtained. This expectation is not completely fulfilled, even though the spreads are smaller than those seen in an earlier series of benchmarks performed some years ago. For all the benchmarks the spread in k-infinity after removing outlying solutions is in excess of 1%; this value would be unacceptable if it was representative of the uncertainty on lattice design calculations. Furthermore, this spread translates into a much larger spread in the plutonium content needed to achieve a given reactivity lifetime.

This means that there is still need for improvement in both methods and basic nuclear data and also for better experimental validation. The spectral shift towards higher energies linked with high plutonium content and/or moderator voidage increases the significance of effects as inelastic slowing down and absorption in the resolved and unresolved resonances. The multirecycling of plutonium emphasises the contribution of higher plutonium isotopes to neutron capture and increases the importance of the uncertainties on the minor actinide cross-sections.

In the absence of experimental data, benchmark exercises are a very useful tool to compare methods and/or basic nuclear data, even if they do not provide the 'true' answer. For future work within the framework of the WPPR, the following issues should be addressed:

- Comparisons between the most recent evaluated data files in the simplest scenario possible;

- In connection with the void reactivity effect benchmark, more realistic conditions should be considered, such as an extension of the voided region beyond the limits of the MOX sub-assembly.

As a final point, it is noted that the design of MOX fuel assemblies needs the use of the best (that normally means the newest) nuclear data bases and the application of very detailed and sophisticated spectra and assembly codes. The use of codes only verified for UO_2 fuel should be avoided.

References

[1]   D. Lutz, A. and W. Bernnat, K. Hesketh, E. Sartori: "Physics of Plutonium Recycling", Volume 2, Benchmark on Plutonium Recycling in PWRs, OECD/NEA (1995).

[2]   D. Lutz, A. and W. Bernnat, K. Hesketh, E. Sartori: "Physics of Plutonium Recycling", Volume 3, Benchmark on Void Reactivity Effect in PWRs, OECD/NEA (1995).

[3]   H. Akie, Y. Ishiguro, H. Takano: "Summary Report on the Comparison of NEACRP Burnup Benchmark Calculations for High Conversion Light Water Reactor Lattices, NEACRP-L-309 (1988).

[4]   W. Bernnat, Y. Ishiguro, E. Sartori, J. Stepanek, M. Takano: "Advances in the Analysis of the NEACRP High Conversion LWR Benchmark Problems, PHYSOR 90, Marseilles, Vol. I, 54 (1990).

[5]   S. Cathalau: "Detailed Comparison of WIMS-6, APOLLO-2, MCNP-4A and SRAC Codes for the PWR Pu Recycling Benchmarks", private communication, 30 November 1994.

Table 3.2.A *k-infinities for Benchmark A*

Contributor Burnup MWd/kg	ANL [1]	BEN	BNFL	CEA	ECN	EDF	HIT	IKE1	IKE2	JAE	PSI1	STU
0.0	1.1324	1.1044	1.1043	1.1334	1.1313	1.1217	1.1396	1.1308	1.1308	1.1336	1.1304	1.1336
10.0	0.0000	1.0400	1.0398	1.0707	1.0746	1.0593	1.0777	1.0688	0.0000	1.0718	1.0686	1.0747
33.0	0.0000	0.9645	0.9645	0.9974	1.0057	0.9863	1.0081	0.9949	0.0000	1.0028	0.9974	1.0055
42.0	0.0000	0.9405	0.9405	0.9716	0.9821	0.9626	0.9833	0.9705	0.0000	0.9799	0.9743	0.9827
50.0	0.0000	0.9208	0.9208	0.9497	0.9622	0.9433	0.9625	0.9507	0.0000	0.9610	0.9554	0.9641

[1] The original ANL result for 300 K is 1.1591 ± 0.0011. It has been converted to the required temperatures using the results of IKE2 for room temperature (1.1586 ± 0.0011) and for benchmark conditions (see Volume 2, [1]).

[2] The original ANL result for 300 K is 1.2117 ± 0.0010. It has been converted to the required temperatures using the results of IKE2 for room temperature (1.2182 ± 0.0011) and for benchmark conditions (see Volume 2, [1]).

Table 3.2.B *k-infinities for Benchmark B*

Contributor Burnup MWd/kg	ANL [2]	BNFL	CEA	ECN	EDF	HIT	IKE1	IKE2	JAE	PSI1	STU
0.0	1.1785	1.1805	1.1896	1.1838	1.1744	1.1926	1.1849	1.1847	1.1872	1.1839	1.1830
10.0	0.0000	1.0923	1.0953	1.0972	1.0824	1.1026	1.0936	0.0000	1.0967	1.0929	1.0947
33.0	0.0000	0.9893	0.9876	0.9953	0.9730	0.9956	0.9851	0.0000	0.9931	0.9870	0.9909
42.0	0.0000	0.9541	0.9496	0.9604	0.9367	0.9586	0.9483	0.0000	0.9579	0.9522	0.9560
50.0	0.0000	0.9254	0.9175	0.9312	0.9070	0.9280	0.9178	0.0000	0.9289	0.9241	0.9269

Contr.	Code	Data	Fuel Type			
			UO$_2$	H-MOX	M-MOX	L-MOX
ANL	VIM	ENDFB5	1.3651±.0010	1.2124±.0012	1.1671±.0011	1.1428±.0006
BEN	LWRWIMS	WIMS	-	1.2054	1.1623	1.1427
CEA3	APOLLO2	JEF22CEA	1.3746	1.2131	1.1711	1.1496
CEA4	ECCO52	JEF22CEA	1.3697	1.2107	1.1674	1.1447
CEA5	APOLLO1	ENDFBCEA	1.3630	1.2092	1.1634	1.1391
CEN	DOT35	MOLBR2	1.3731	1.2255	1.1952	1.1835
ECN	MCNP4	JEF22	1.3746±.0005	1.2166±.0007	1.1725±.0008	1.1504±.0009
ENEA	MCNP4	JEF1	1.3608±.0006	1.2131±.0008	1.1676±.0008	1.1435±.0008
HIT	VMONT	JENDL2	1.3582	1.2237	1.1799	1.1549
IKE1	CGM	JEF1	1.3633	1.2137	1.1710	1.1488
IKE2	MCNP4	JEF22	1.3726±.0003	-	-	-
JAE1	SRAC	JENDL31	1.3618	1.2213	1.1782	1.1544
JAE2	MVP	JENDL31	1.3622±.0003	1.2185±.0004	1.1757±.0004	1.1533±.0004
JAE3	SRACPIK	JENDL31	1.3618	1.2241	1.1782	1.1544
JAE4	MOSRA	JENDL31	1.3618	1.2213	1.1782	1.1544
TOS	MCNP42	JENDL31	1.3609±.0010	1.2165±.0011	1.1715±.0009	1.1501±.0012
Average			1.3656	1.2163	1.1733	1.1511

Table 3.4 k-infinity pin cell moderated

Contr.	Code	Data	Fuel Type			
			UO$_2$	H-MOX	M-MOX	L-MOX
ANL	VIM	ENDFB5	0.6215±.0006	1.2850±.0007	1.0380±.0006	0.7616±.0011
BEN	LWRWIMS	WIMS	-	1.2592	1.0217	0.7574
CEA3	APOLLO2	JEF22CEA	0.6500	1.2879	1.0475	0.7766
CEA4	ECCO52	JEF22CEA	0.6444	1.2850	1.0441	0.7738
CEA5	APOLLO1	ENDFBCEA	0.6234	1.2860	1.0398	0.7637
CEN	DOT35	MOLBR2	0.6211	1.2465	1.0051	0.7379
ECN	MCNP4	JEF22	0.6380±.0006	1.2863±.0008	1.0427±.0004	0.7696±.0005
ENEA	MCNP4	JEF1	-	1.2668±.0011	1.0216±.0013	0.7495±.0018
HIT	VMONT	JENDL2	0.6211	1.2781	1.0324	0.7595
IKE1	CGM	JEF1	0.6212	1.2712	1.0284	0.7573
IKE2	MCNP4	JEF22	-	-	-	-
JAE1	SRAC	JENDL31	0.6264	1.2741	1.0331	0.7650
JAE2	MVP	JENDL31	0.6228±.0003	1.2689±.0004	1.0280±.0003	0.7598±.0005
JAE3	SRACPIK	JENDL31	0.6264	1.2741	1.0331	0.7651
JAE4	MOSRA	JENDL31	0.6264	1.2741	1.0331	0.7650
TOS	MCNP42	JENDL31	0.6236±.0013	1.2783±.0013	1.0369±.0013	0.7657±.0011
Average			0.6282	1.2748	1.0324	0.7618

Table 3.5 k-infinity pin cell voided

Contr.	Code	Data	Fuel Type			
			UO$_2$	H-MOX	M-MOX	L-MOX
ANL	VIM	ENDFB5	1.3653±.0002	1.3428±.0002	1.3391±.0002	1.3382±.0002
BEN	LWRWIMS	WIMS	-	1.3452	1.3418	1.3415
CEA3	APOLLO2	JEF22CEA	1.3745	1.3508	1.3472	1.3464
CEA4	ECCO52	JEF22CEA	1.3697	1.3464	1.3428	1.3417
CEA5	APOLLO1	ENDFBCEA	1.3630	1.3403	1.3363	1.3353
CEN	DOT35	MOLBR2	1.3773	1.3561	1.3537	1.3538
ECN	MCNP4	JEF22	1.3773±.0010	1.3516±.0007	1.3495±.0007	1.3481±.0008
ENEA	MCNP4	JEF1	1.3600±.0007	1.3228±.0007	1.3201±.0008	1.3198±.0007
HIT	VMONT	JENDL2	1.3577	1.3376	1.3341	1.3324
IKE1	CGM	JEF1	1.3657	1.3438	1.3387	1.3392
IKE2	MCNP4	JEF22	1.3726±.0003	1.3487±.0003	1.3446±.0003	1.3440±.0003
JAE1	SRAC	JENDL31	1.3617	1.3404	1.3367	1.3357
JAE2	MVP	JENDL31	1.3629±.0003	1.3416±.0003	1.3383±.0003	1.3374±.0003
JAE3	SRACPIK	JENDL31	1.3618	1.3411	1.3373	1.3361
JAE4	MOSRA	JENDL31	1.3615	1.3403	1.3367	1.3356
TOS	MCNP42	JENDL31	1.3601±.0010	1.3382±.0009	1.3364±.0009	1.3334±.0009
Average			1.3661	1.3430	1.3396	1.3387

Table 3.6 *k-infinity macrocell moderated*

Contr.	Code	Data	Fuel Type			
			UO$_2$	H-MOX	M-MOX	L-MOX
ANL	VIM	ENDFB5	1.3508±.0002	1.3481±.0002	1.3434±.0002	1.3398±.0002
BEN	LWRWIMS	WIMS	-	1.3513	1.3469	1.3440
CEA3	APOLLO2	JEF22CEA	1.3545	1.3542	1.3486	1.3452
CEA4	ECCO52	JEF22CEA	1.3557	1.3507	1.3461	1.3430
CEA5	APOLLO1	ENDFBCEA	1.3492	1.3460	1.3410	1.3375
CEN	DOT35	MOLBR2	1.3609	1.3584	1.3537	1.3507
ECN	MCNP4	JEF22	1.3618±.0005	1.3583±.0005	1.3547±.0008	1.3505±.0007
ENEA	MCNP4	JEF1	1.3454	1.3248±.0007	1.3196±.0008	1.3155±.0007
HIT	VMONT	JENDL2	1.3434	1.3419	1.3371	1.3330
IKE1	CGM	JEF1	1.3501	1.3455	1.3408	1.3377
IKE2	MCNP4	JEF22	1.3588±.0003	1.3549±.0003	1.3502±.0003	1.3466±.0003
JAE1	SRAC	JENDL31	1.3460	1.3408	1.3363	1.3332
JAE2	MVP	JENDL31	1.3474±.0003	1.3457±.0003	1.3415±.0004	1.3372±.0003
JAE3	SRACPIK	JENDL31	1.3468	1.3450	1.3401	1.3364
JAE4	MOSRA	JENDL31	1.3459	1.3428	1.3379	1.3344
TOS	MCNP42	JENDL31	1.3450±.0009	1.3422±.0009	1.3390±.0009	1.3356±.0009
Average			1.3508	1.3469	1.3423	1.3388

Table 3.7 *k-infinity macrocell voided*

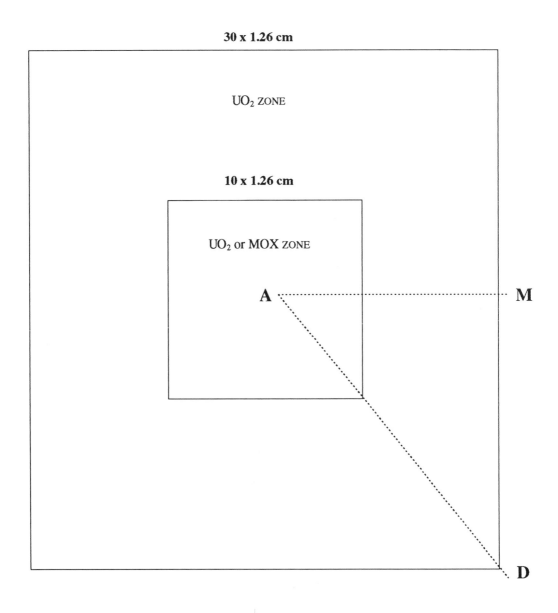

Figure 3.1
Supercell configuration of a 30x30 array of PWR fuel cells for the void reactivity effect benchmark

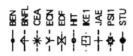

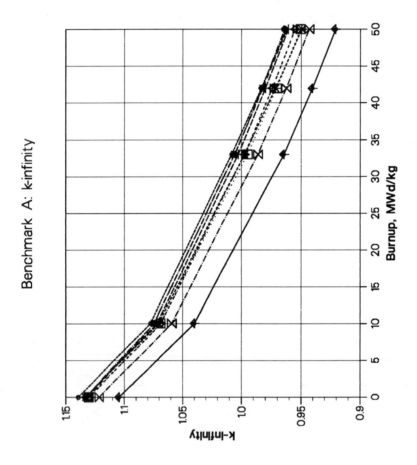

Figure 3.2.A k-infinity benchmark A

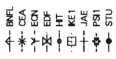

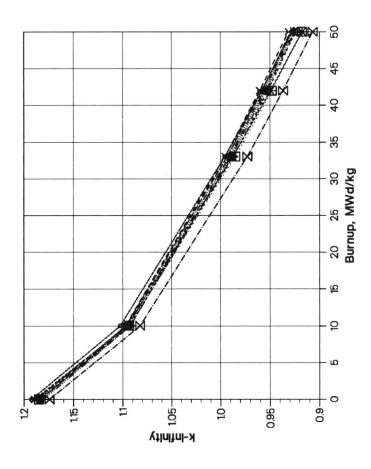

Figure 3.2.B k-infinity benchmark B

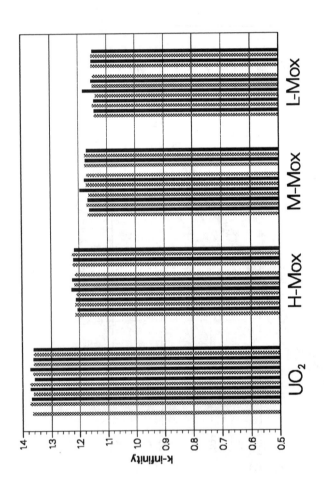

Figure 3.3 k-infinity pin cell moderated

88

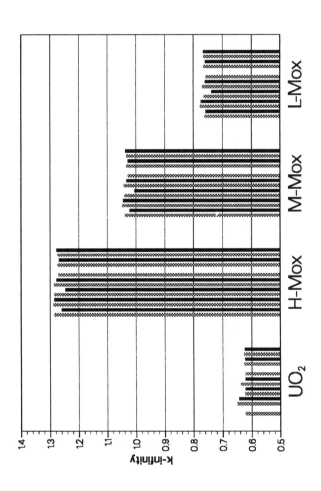

Figure 3.4 k-infinity pin cell voided

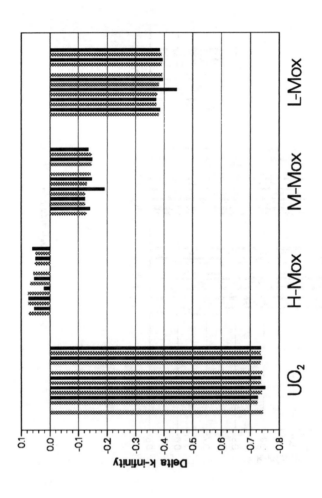

Figure 3.5 Δk-void infinite lattice

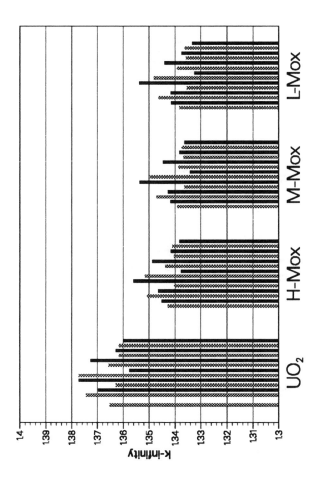

Figure 3.6 k-infinity macrocell moderated

91

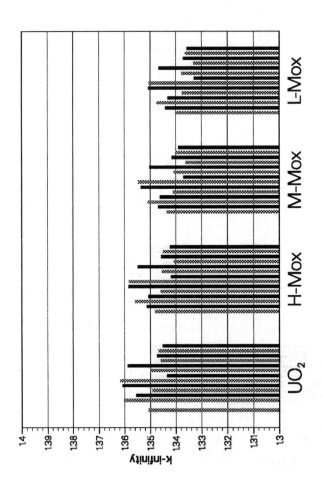

Figure 3.7 k-infinity macrocell voided

92

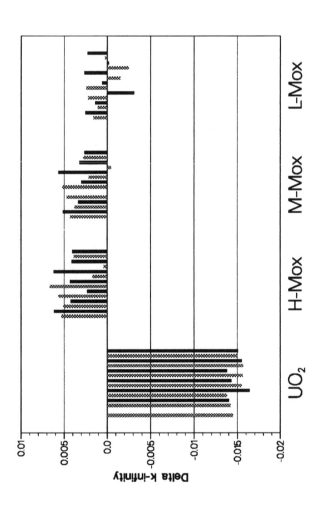

Figure 3.8 Δk-void macrocell

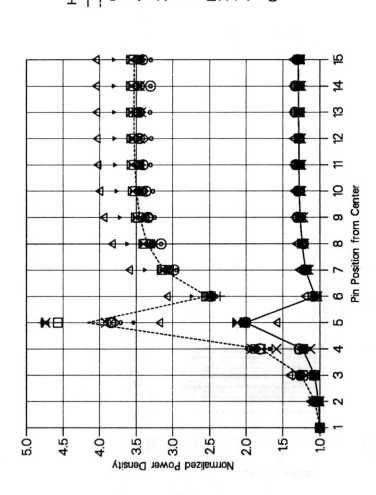

Figure 3.9 Fission density traverse diagonal H-MOX

Chapter 4

PLUTONIUM RECYCLING AND WASTE/RADIOTOXICITY REDUCTION

4.1 Introduction

A major concern in shaping the future of the nuclear power industry is the ecologically acceptable management of waste. As said previously, even in countries which have developed extensive nuclear power generation facilities, no unique strategy has emerged for the management of spent thermal reactor fuel. General public acceptance of final disposal of high-level radioactive waste is increasingly related to the awareness of the very long-term environmental issues associated with the radiotoxicity of components of the waste – specifically long-lived fission products and long-lived man-made transuranic elements (transuranics).

It is then very relevant to address the issue of radiotoxicity flows in considering plutonium recycling. In what follows we will recall different scenarios for the use of plutonium and the transuranic element inventories, as discharged from present thermal reactors. Flows and working inventories of radiotoxicity will be discussed next and, finally, some relevant issues related to the recycle of transuranics and the potential for waste volume reduction.

4.2 Different scenarios for the use of plutonium

Different scenarios (which are not necessarily exclusive) can be envisioned for the management of spent fuel from thermal reactors (Figures 4.1 to 4.3):

- Direct and final disposal in an engineered repository;

- Long term storage, leaving open the option for reprocessing at later times;

- Immediate reprocessing to separate the fuel (recovered uranium and plutonium and later on other transuranic elements) from the fission products for re-fabrication into fuel for reactors.

In this last case different scenarios can be developed according to whether the transuranic elements other than plutonium are treated as wastes or separated from fission products and recycled in reactors for transmutation. The reactors in which plutonium (or eventually other transuranic elements) is recycled can be thermal or fast reactors.

Among thermal reactors, LWRs and HWRs have been used or considered [1, 2, 3, 4]. LWRs with different moderator to fuel ratios can be envisaged [5, 6]. Among fast reactors, different types of fuel can be considered for the core (oxide, metal, possibly nitride) with their specific fuel cycles [7, 8] and

95

with specific values of the conversion ratio, according to the prevalent function assigned to this type of reactor (breeding or burning plutonium [9]). In Figure 4.4 three possible scenarios using MOX recycling in PWRs are illustrated [5], with or without multiple recycling and with or without the use of MOX-fuelled fast reactors.

In Figure 4.5 a rationale for the use of plutonium in fast reactors is illustrated, showing both the PUREX reprocessing strategy and pyrometallurgy-based option. This last scheme forms the basis of the benchmarks on Pu recycling in fast reactors (see Chapter 5).

4.3 Thermal reactor discharged transuranic elements inventories

The fuel throughputs differ for the different types of thermal reactors, and, for each reactor, depending on annual feed rates (and load factors) and discharge burnups. Some typical examples are given in Table 4.1.

However, from the physics point of view, when considering plutonium, it is of interest to specify the five Pu isotopes, which have different physics properties. As an example, Table 4.2 provides the isotopic composition of discharged Pu according to the burnup of UO_2 fuel in a PWR.

As has been indicated, the problem of the use of plutonium cannot be viewed as independent from the management of the other transuranics, if the goals of optimum use of resources are to be coupled to the reduction of wastes.

It is then necessary to consider the inventories of the other transuranic elements present in the spent fuel. These elements have different physics characteristics and may be classified according to four families:

- *Neptunium*: The isotope 237 is mainly formed from U-235 (by neutron capture and beta decay, and, to a lesser extent, from U-238 by (n,2n) reaction. Its short-term production is independent of the eventual recycling of Pu. Subsequent to discharge neptunium is formed from decay of Pu-241 to Am-241 to Np-237. Its half-life is extremely long;

- *Americium 241*, generated by the beta decay of Pu-241 which has a relatively short decay period (~ 14 years): Its production is therefore greater in the fuel cycle than in reactors. As a strong alpha and gamma emitter, its presence penalises reprocessing and MOX fuel fabrication; since its production is mainly related to delays between reprocessing and fabrication, early fabrication into MOX fuel has beneficial consequences on its management since in current fuel cycle downstream designs, it is in part associated with plutonium (after plutonium separation during reprocessing), in part isolated (in the case of removal of americium from plutonium), and in part mixed with other americium isotopes (contained in current reprocessing waste);

- *Am-242 and Am-243*, produced by irradiation of isotopes higher than Pu: Am-242 by neutron capture in Am-241 (and hence the quantities are negligible); Am-243 created from beta decay of Pu-243, which has a very short half-life of ~ 5 hours. They therefore depend on the Pu grade and the irradiation conditions;

- *Curium and heavier nuclides*, also produced from successive captures and beta decays from plutonium through americium. Spent fuel comprises less than 0.01 % of curium (Cm-242, Cm-244 and Cm-245). The half-lives of Cm-242 and Cm-244 are short but their presence even in small amounts gives rise to strong radiation and heat sources.

Table 4.3 gives some typical values for minor actinides discharged from UO_2 fuel with various burnups in PWRs [5].

4.4 Flows and working inventories of radiotoxicity

As regards the nuclear waste management, radiotoxicity flows per unit of energy benefit derived from fission reactors and working inventories of radiotoxicity per unit of installed capacity are of key interest. One must consider the inflow, outflow and working inventory of radiotoxicity from the macrosystem shown in Figure 4.6.

Because the radiotoxicity varies with the age of the material subsequent to discharge, it is helpful to divide the radiotoxicity contributions into several categories:

- Those inflows and outflows from the Figure 4.6 macrosystem which apply for geologic time scales, including:

  - the radiotoxicity of the uranium ore as an inflow,
  - the radiotoxicity of the long-lived fission products as an outflow, and,
  - the radiotoxicity of any unused uranium as an outflow;

- Those which apply for societal times scales (~ 500 years or less):

  - this includes fission products with half lives less than several hundred years;

  and finally:

- Those radiotoxicity flows associated with the man-made transuranics which are created as a neutronic consequence of subjecting the uranium to interactions with neutrons.

Sekimoto [10] has considered a macrosystem such as Figure 4.6 in the steady-state. He shows that the increase in the earth's radiotoxicity burden due to long-lived fission products per unit of energy is a little more than cancelled by the reduction in the earth's radiotoxicity burden which occurs because the uranium ore is removed from its crust and fissioned to create the energy in the first place. Figure 4.7, [10] shows the radiotoxicity flows associated with the uranium ore required for one year's operation of a 3 GWth fast reactor and the fission products produced. Three cases are considered :

- Only the uranium is withdrawn from the earth's crust (the practical case);

97

- The uranium, and the thorium which is in secular equilibrium with it is removed, and finally;

- The radium as well as thorium which is in secular equilibrium is removed with the uranium.

In all cases, on geologic time scales, a radiotoxicity of a little over 10^6 Annual Limit on Intake (ALI) units are removed from the earth's crust in order to supply the uranium needed for one year's worth of 3 GWth reactor operation. Figure 4.7 [10] shows that the long-lived fission products created and returned to the earth's crust after fissioning that uranium is a little less than 10^6 ALI units. Thus, the year's worth of 3 GWth energy benefit would be "free" in terms of long-term ecological impact if the short-lived fission products and the unconsummated actinides (uranium and transuranics) are not returned to the earth.

If one wished to exploit the opportunity for the benefit of fission energy free of net long-term ecological impact (*as measured by radiotoxicity legacy for geologic times*), one would have to find a way to avoid returning actinides and short-lived fission products from the nuclear power macrosystem (see Figure 4.6) to the earth's crust. The approaches for avoiding their return are limited in number:

a) For the actinides: to retain them in working inventory until they are fissioned is the only option available;

this applies to:

- any unused uranium, and
- any man-made transuranics;

b) For the short-lived fission products two options can be considered:

- sequester them in a storage inventory (such as a geological repository) for several hundred years until they have decayed to stable isotopes, or;
- transmute them into stable isotopes.

Whichever of the above approaches is taken, one can view them as a strategy of "hold-up actinides in working inventory until such time as their energy content has been extracted". A quantification – actinides through related benchmarks – of options a) and b) is given in Chapter 5.

4.5 Sequestering of short-lived fission products

Considering first the short-lived fission products, which comprise about 3 w/o of LWR spent fuel, Figure 4.8 [1] shows that within three hundred years, all short-lived fission products will have decayed if not to stable isotopes, at least to a level below that of the long-lived fission products. For short times after discharge (i.e., several decades), however, the short-lived fission products are three orders of magnitude higher in toxicity than are the long-lived fission products and comprise a significant fraction of the radioisotopic hazard of spent nuclear fuel.

---

[1] Figure 4.8 applies to LWR spent fuel after 33 MWd/kg average burnups.

Slessarev, et al. [11] have shown that their total transmutation into stable isotopes in a reactor would require more neutrons than are physically available. If separation technologies were developed, transmutation of specific short-lived fission products would be neutronically feasible. Alternately, their rapid decay vis-à-vis geologic time scales makes sequestering in an interim storage working inventory an attractive option. Containers, engineered structures, warning signs, and even societal institutions for stewardship – all capable of lasting several hundred years – are within the realm of demonstrated human achievement. Stewardship of this sequestered interim storage inventory will comprise a cost per unit of energy benefit which is only just now being quantified.

4.6 Recycle of unused uranium

The unused uranium represents 96 to 97 w/o of LWR spent fuel. Moreover, the tails from the enrichment process also comprise a very large mass of unused uranium. It is clear from Figure 4.7 that if this unused uranium were returned to the earth's crust, it would essentially replace that which had been removed in the first place, and the radiotoxicity from the long-lived fission products would then constitute a small but none-the-less net positive ecological impact on geologic time scale radiotoxicity as the price of having extracted the fission energy.

To the extent that the unused uranium can be held in working inventory for later recycle to ultimate destruction by fission, its radiotoxicity will not become a contributor to geologic time scale radiotoxicity return to the earth's crust as a result of deriving the energy benefit from fission. Moreover, because the uranium can be chemically separated from the transuranics and fission products, it can be put into a relatively benign form for interim storage prior to eventual fissioning.

4.7 Recycle of transuranics

Considering next the radiotoxicity due to the man-made transuranic elements (which comprise about 1 w/o of the actinide content of LWR discharge fuel), Figure 4.8 shows that the initial radiotoxicity rivals that of the short-lived fission products and that the very long-term radiotoxicity exceeds that of the long-lived fission products. This is because the half lives of many of the actinides are of the geologic time range. In the case of the transuranics then, their return to the earth's crust would more than double the long-lived fission product radiotoxicity burden applicable for geological times. Alternately, if they are not returned to the earth's crust but instead are retained in working inventory, recycled and fissioned, they can be consumed in tens to hundreds of years and will produce fission products which add to the short-lived fission product inventory destined for interim storage for several hundred years. They will also contribute to the long-lived fission product radiotoxicity returned to the earth's crust – but still sum to a reduced level relative to that drawn from the earth with the original uranium ore.

To the degree that the technology for recycle allows a leakage of transuranics into the waste product for return to the earth's crust vis-à-vis retention in working inventory, the transuranics will add incrementally to the geologic time scale radiotoxicity of the long-lived fission products. This leakage is a function of recycle technology and is one of the focal points of the Working Party's benchmarking exercise for fast reactors (see Chapter 5).

The extent to which recycled transuranics can be fissioned completely is a function of the neutron spectrum of the reactor, and again is a focal point of the benchmark exercises; however, it is well known that the neutron interaction properties of the actinides dictate that while thermal recycle will work for a while, a fast spectrum will be ultimately required during the final stages of the multiple recycle.

4.8 Neutron interaction properties of plutonium and other transuranic elements

In the present section the major neutron interaction characteristics of Pu isotopes and other transuranics will be given with respect to a) the neutron economy in a reactor and b) the fission/capture cross-section ratio, related to their transmutation potential, in view of waste/radiotoxicity reduction during recycling.

4.8.1 The neutron economy

A way to evaluate the impact on the neutron economy of the introduction in a neutron flux of a particular isotope i, is to establish the neutron consumption (or production) D_i per fission [11], in the process of full utilisation of that isotope, down to complete fission (i.e. of the initial isotope and all the isotopes eventually formed from it by neutron capture, etc.).

For a fuel made up by I isotopes, the fuel neutron consumption (or production) is defined as follows:

$$D_{Fuel} = \sum_{i=1,I} \varepsilon_i \, D_i$$

(ε_i = fraction of fissions due to isotope i).

The condition of positive core neutron economy is given by:

$$-D_{Fuel} + (CM + L) \geq 0$$

where CM are parasitic captures (structures, fission products, etc.) per fission and L the neutron leakage per fission.

D_i varies from isotope to isotope, according to the neutron spectrum. Since a specific reactor core type is associated with a specific neutron spectrum and flux level, it is possible to evaluate the impact of the introduction of different isotopes in that specific reactor type; D_i values are given in Table 4.4 for Pu isotopes and minor actinides in different reactor types. Since the CM+L term is approximately 0.3 neutrons/fission in most reactor types, values of Table 4.4 can help to evaluate the impact of the different Pu isotopes on the neutron economy of a specific reactor type, chosen for recycling. The greater ease with which total transuranic composition can be achieved in a fast versus a thermal spectrum reactor is reflected in the much more negative value of D_{TRU} displayed in the last column of Table 4.4.

4.8.2 The fission/capture ratios

If Pu recycling in reactors is to be optimised both from the core performance point of view, and also with the objective of reducing the wastes and their radiotoxicity, then this means essentially minimising the production of heavier isotopes.

This last objective is clearly related, from a physics point of view to the fission/capture cross-section ratios; the more favourably fission competes with capture, the more efficiently one avoids the production of heavier isotopes.

Table 4.5 gives typical fission/capture values for actinides in different neutron spectra. These ratios are maximum for hard fast neutron spectra. A lower minor actinide production is then expected from Pu recycling in a fast reactor with respect to a thermal (e.g. PWR) reactor.

However it would be possible to envisage optimising the situation for thermal reactors. It has been indicated that in the case of PWRs, taken again as example, decreasing of the fuel-to-moderator ratio reduces the buildup of minor actinides (see Table 4.6). This shows that a reactor with a high moderating ratio, although far from equalling the performance of a fast spectrum reactor, is better than a conventional PWR at reducing actinides at the first Pu recycling. This tendency results from the lower initial Pu content, despite the lower fission/capture cross-section ratio values in this type of spectrum. In fact, the tendency obtained at the first recycling can be changed if multiple recycling is considered (see Table 4.7) even if at each cycle the Pu coming from MOX spent fuels is diluted with fresh Pu coming from UO_2 spent fuels.

The production of minor actinides in MOX fuelled LWRs is also sensitive to the burnup and to the initial quality of the recycled plutonium. Table 4.8 gives some typical examples which show a clear trend of an increase in the heaviest nuclei buildup with burnup.

4.9 Conclusions

The objective of waste/radiotoxicity reduction has to be accounted for when evaluating different plutonium utilisation strategies.

From the physics point of view, the major issues relate to the characteristics of the neutron spectra of the different reactor types. There is a general consensus acknowledging the favourable characteristics of fast neutron spectra. The analysis in Chapter 5 is significant in this respect.

Multirecycling of plutonium in thermal reactors (see Chapter 2) can produce a significant increase of minor actinides and then an increase of the potential radiotoxicity source. This is a practical near term option for extracting further energy from the actinides and it does not preclude the subsequent further recycle in a fast reactor to reverse the trend of toxicity buildup. Moreover, what matters in the long term is what radiotoxicity is actually returned to the earth's crust.

Some optimisation could be envisaged (i.e. with respect to the light water moderator-to-fuel ratio) but further work would be needed to identify limitations arising from minor actinide buildup during plutonium multirecycling. However, in order to keep the ratio of minor actinides/plutonium as low as possible and achieve as high a burnup as possible, self sufficient fast reactors are still the best option.

References

[1]   G.J. Schlosser, S. Winnik: "Thermal Recycling of Plutonium and Uranium in the Federal Republic of Germany" - IAEA-SM-294/33.

[2]   M. Rome et al.: "Plutonium Reload Experience in French Pressurized Water Reactors" – Nucl. Technology, Vol. 94 (1991).

[3]   P. G. Borzat, I. J. Hastings, A. Celli: "Recycling in CANDU 5 % Uranium and/or Plutonium from Spent LWR Fuel" – IAEA, IWGFPT/35.

[4]   S. Sawai et al.: "Characteristics of Pu Utilisation in the Heavy-Water Moderated Boiling-Light-Water-Cooled Reactor ATR" – Nucl. Eng. and Design 125 (1991).

[5]   J. Bouchard: "The Effect of Fuel Cycle Strategies on the Choice of Next-Generation Reactors" – ENS TOPNOX'93, The Hague, April 25-28, 1993.

[6]   K. Kanda et al.: "Effective Utilisation of Uranium and Plutonium in LWR Cores" – ANS Winter Meeting, Chicago 1992.

[7]   H. H. Hennies, J. Leduc, S. C. Goddard: "Development of Fast Reactors in Europe", paper 1-2
       M. Hori, A. Takeda: "Development of the FBR as the Leading Energy Source for Japan", paper 1-5
       Proc. Int. Conf. on Fast Reactors and Related Fuel Cycles (FR '91), Kyoto, 28 October - 1 November 1991.

[8]   C. E. Till and Y. I. Chang: "The Integral Fast Reactor Concept", Proc. Am. Power Conf. 48, 688 (1986)
       C. E. Till and Y. I. Chang: "The Integral Fast Reactor", Advances in Nuclear Science and Technology, 20, 127 (1988).

[9]   J. Rouault et al.: "Physics of Pu Burning in Fast Reactors" – Proc. ANS Topical Meeting on Reactor Physics, Knoxville (1994).

[10]   H. Sekimoto: "Physics of Future Equilibrium State of Nuclear Energy Utilization" – Proc. Intl. Conf. on Reactor Physics and Reactor Computations, Tel-Aviv, 23-26, Jan. 94. Y. Ronen and E. Elias, eds, Ben Gurion University of the Negrev Press (1994).

[11]   M. Salvatores, I. Slessarev, M. Uematsu, Nucl. Sci. Eng. 116, 1-18 (1994).

[12]   L. Koch: "Formation and Recycling of Minor Actinides in Nuclear Power Stations" – Handbook on the Physics and Chemistry of Actinides, Volume 4, Chapter 9,
       A. J. Freeman and C. Keller, eds, Elsevier Science Publishers B.V. (1986).

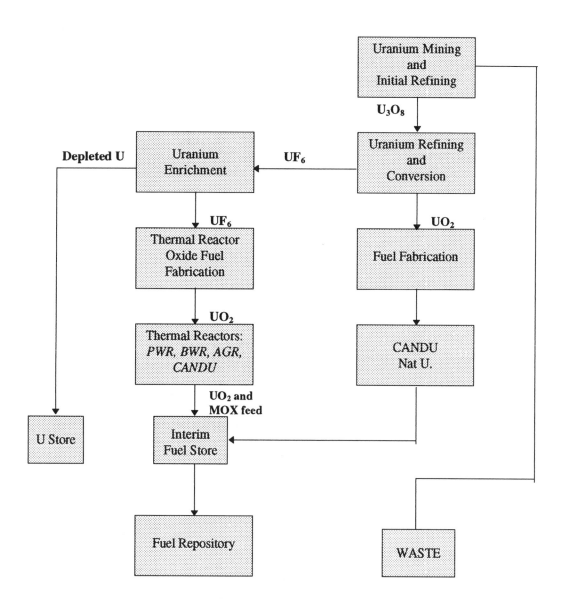

*Figure 4.1 **Thermal reactor fuel cycle with spent fuel storage and disposal***

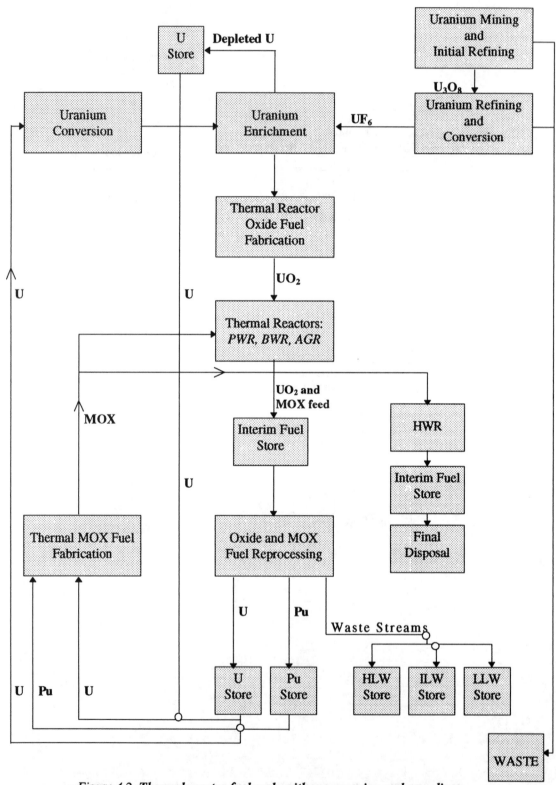

*Figure 4.2 **Thermal reactor fuel cycle with reprocessing and recycling***

104

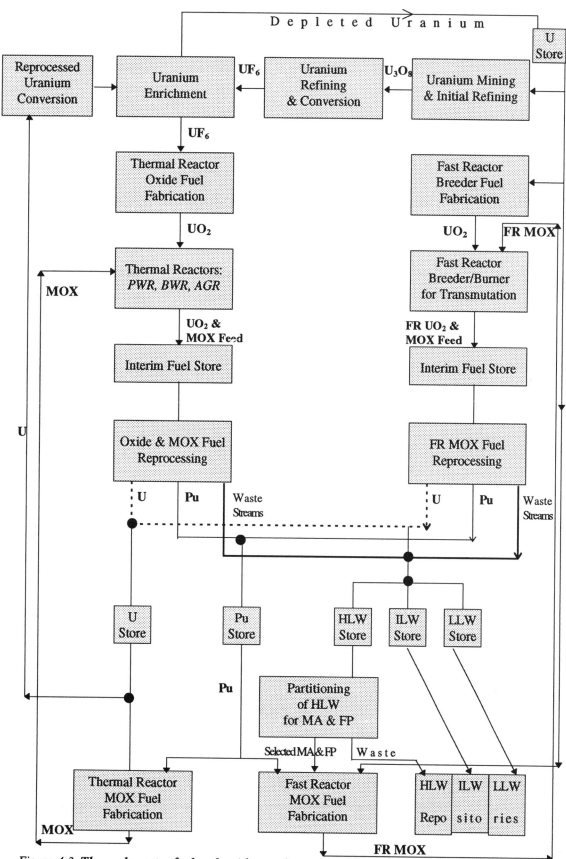

*Figure 4.3 **Thermal reactor fuel cycle with recycling and actinide burning in fast reactors***

SCENARIO 1:   Multiple recycling in PWRs

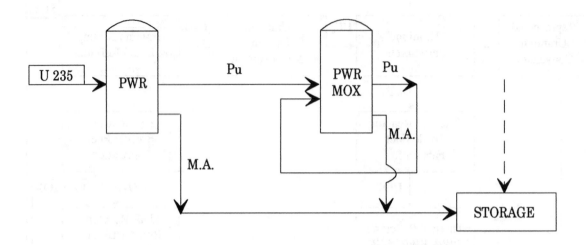

SCENARIO 2:   Single recycling in PWRs followed by multiple recycling in a burner

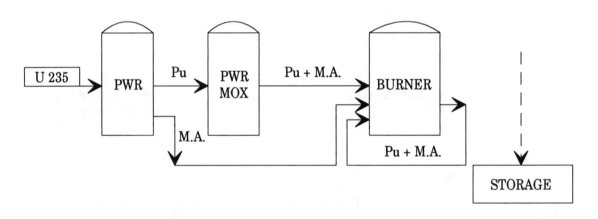

SCENARIO 3:   Optimised multiple recycling in PWRs followed by multiple recycling in a burner

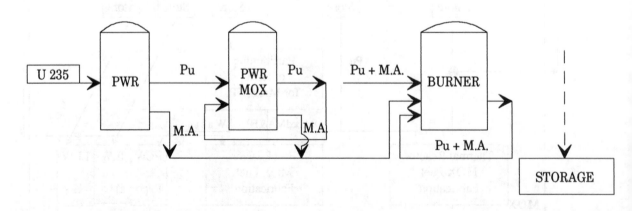

Figure 4.4 Recycling scenarios for MOX fuel

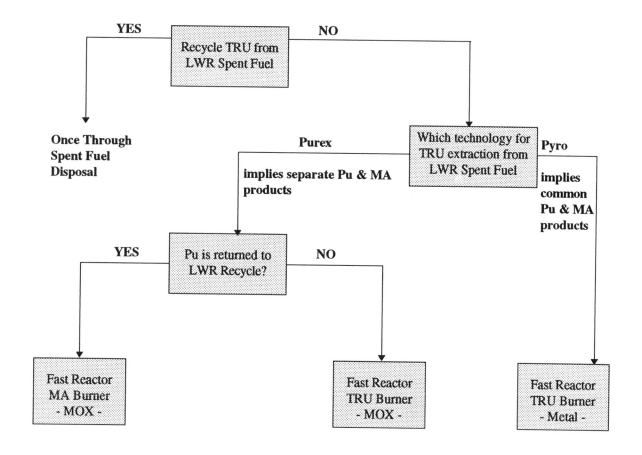

Figure 4.5 Rationale for the use of plutonium in fast reactors

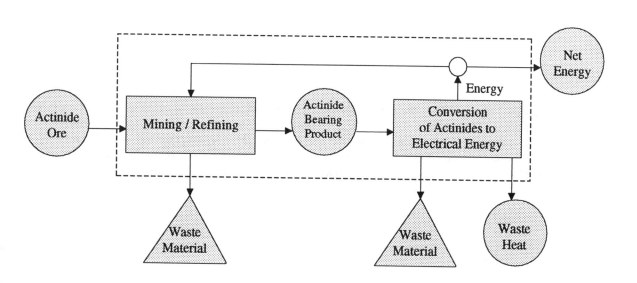

Figure 4.6 Macro system view of fission - induced energy extraction from earth's endowment of actinides

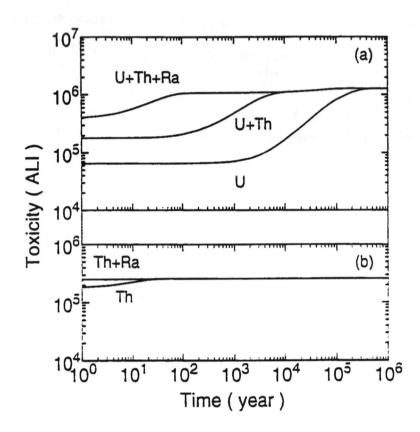

Figure 4.7 **Expected change along the time of the incinerated toxicity measured by Annual Limit on Intake of (a) natural uranium (uranium only, uranium + thorium and uranium + thorium + radium) fed to the 8 GWth soft-spectrum fast reactor per year and (b) thorium (thorium only and thorium + radium) fed to the 3 GWth hard-spectrum thermal reactor per year. The amounts of the additionally charged nuclides are the equilibrium amounts to each parent (uranium or thorium)** [10].

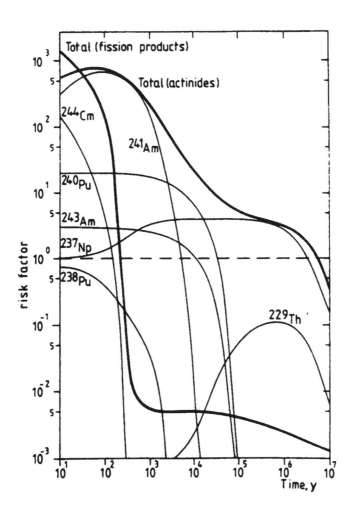

Figure 4.8
Time dependence of the risk (hazard) factor for wastes from spent fuel, stored without reprocessing [12].

REACTOR TYPE	MAGNOX	AGR	LWR	LWR	CANDU
Fuel type	Metal	Oxide	UO_2	MOX (100%)	Oxide
Burnup (GWd/t or MWd/kg)	5	20	33	33	7,9
Pu (kg/t)	3.8	5.0	9.3	- 15	3.8
MA (kg/t)		0.18	0.9	4	0.05
FP (kg/t)	5	19	34	34	7.9

Table 4.1 Discharged fuel content for different types of reactors

BURNUP (MWd/kg)	Pu-238	Pu-239	Pu-240	Pu-241	Pu-242
7.9	1.2	67.3	25.5	4.5	1.5
33	1.8	57.9	22.5	11.1	5.6
42	2.7	54.5	22.8	11.7	7.0
55	4.0	50.4	23.0	12.3	9.1
65	4.8	47.5	23.8	12.1	10.5

Table 4.2 Plutonium isotopic composition from spent UO_2 fuel in a PWR (after 5 years' decay)

BURNUP (MWd/kg)	INITIAL ENRICHMENT	TOTAL Pu	Np-237	Am-241	Am-243	Cm-244	Cm-245
7.9	0.72 %	67	0.67	0.87	0.05	.003	.00003
43	3.70 %	34	1.81	0.58	0.49	0.14	0.01
55	4.50 %	28	1.70	0.47	0.56	0.22	0.02
65	4.95 %	25	1.90	0.48	0.71	0.34	0.03

Table 4.3 Pu and minor actinides from spent UO_2 fuel in a PWR (kg/TWhe) (after 5 years' decay)

	NEUTRON SPECTRA				
	FR - oxide fuel		LWR - thermal		Super-thermal
A - families $\Downarrow$	$\Rightarrow$ ϕ (n/cm^{2}s) 10^{15}	10^{17}	10^{14}	10^{16}	10^{16}
Pu-238	- 1.36	- 1.49	0.17	0.042	- 0.13
Pu-239	- 1.46	- 1.51	- 0.67	- 0.79	- 1.07
Pu-242	- 0.44	- 0.75	1.76	1.10	1.12
Np-237	- 0.59	- 0.72	1.12	0.53	- 0.46
Am-243	- 0.60	- 1.07	0.82	0.16	0.21
Cm-243	- 2.13	- 2.26	- 1.90	- 2.04	- 1.63
Cm-244	- 1.39	- 1.92	- 0.15	- 0.53	- 0.48
D_{TRU}	- 1.19	- 1.33	- 0.077	- 0.38	- 0.58

Table 4.4
D_A - Values for some important TRU-nuclei and for standard LWR discharge in different neutron spectra and neutron fluxes ϕ (n/cm^{2}s)

ISOTOPE	LIGHT WATER REACTOR MOX FUEL	FAST NEUTRON REACTOR
Np 237	0.03	0.19
Pu-238	0.20	1.60
Pu-239	1.90	3.00
Pu-240	0.02	0.60
Pu-241	2.90	5.00
Pu-242	0.04	0.33
Am-241	0.02	0.15
Am-242	5.00	4.60
Am-243	0.01	0.12
Cm-243	8.70	16.50
Cm-244	0.12	0.57
Cm-245	7.00	9.00

Table 4.5 Fission / capture cross-section ratio for some actinides

	Np-237	Am-241	Am-243	Cm-244	Cm-245
Under Moderated PWR :MOX 42 MWd/kg	0.7	13.7	5.6	2.5	0.5
Standard PWR : "	0.6	8.2	4.3	2.0	0.3
Highly Moderated PWR : "	0.5	4.7	4.2	1.7	0.15
Fast Neutron Reactor : 125 MWd/kg	0.3	3.6	1.4	0.31	0.03

(after 5 years decay)

Table 4.6
Minor actinides production in PWRs and FBRs (kg/TWhe: single recycling of plutonium in PWRs)

	$V_m/V_f = 3.0$			$V_m/V_f = 2.0$			$V_m/V_f = 1.1$		
	cycle=1	= 2	= 3	cycle=1	= 2	= 3	cycle=1	= 2	= 3
Pu fissile content (%)	4.2	4.8	5.2	6.0	6.8	7.3	8.8	9.0	9.1
ΔPu (kg/TWhe)	- 68	- 78	- 83	- 64	- 71	- 76	- 54	- 57	- 59
ΔMA (kg/TWhe)	+ 11	+ 15.8	+ 18.6	+ 16.4	+ 20.8	+ 23.2	+ 21.8	+ 23.0	+ 23.8
$|\Delta$MA/ΔPu$|$	0.16	0.20	0.22	0.26	0.29	0.30	0.40	0.40	0.40

Case 1: Multirecycling with dilution

	$V_m/V_f = 3.0$			$V_m/V_f = 2.0$			$V_m/V_f = 1.1$		
	cycle=1	= 2	= 3	cycle=1	= 2	= 3	cycle=1	= 2	= 3
Pu fissile content (%)	4.2	7.4	9.1	6.0	7.6	8.6	8.8	9.2	9.3
ΔPu (kg/TWhe)	- 68	- 104	- 122	- 64	- 79	- 90	- 54	- 59	- 63
ΔMA (kg/TWhe)	+ 11	+ 32.5	+ 45.	+ 16.4	+ 25.6	+ 32.2	+ 21.8	+ 24	+ 25.9
$|\Delta$MA/ΔPu$|$	0.16	0.31	0.37	0.26	0.32	0.36	0.40	0.41	0.42

Case 2: Multirecycling with no dilution

Table 4.7
**Pu consumption and minor actinide (MA) production during multi-recycling in PWRs
with different moderator-to-fuel ratios (burnup = 55 MWd/kg)**

RECYCLING	BURNUP (MWd/kg)	Np-237	Am-241	Am-243	Cm-244	Cm-245
MOX 1st	42	0.6	8.2	4.3	2.0	0.3
MOX 1st	55	0.5	7.5	4.9	2.7	0.5
MOX 1st	65	0.5	8.6	5.7	3.3	0.7
MOX 2nd	65	0.5	12.5	7.3	3.7	0.8
MOX 3rd	65t	0.5	13.0	8.4	4.0	0.9

Table 4.8
Minor actinides production in PWRs – variations with burnup and successive recycling (kg/TWhe)

Chapter 5

PLUTONIUM RECYCLING IN FAST REACTORS

5.1 Physics of plutonium recycle

As discussed in the previous chapter, in a macrosystem sense, the nuclear power enterprise is one in which uranium ore is mined and refined at some expense of energy – thereby producing both a uranium product and a waste stream of milling by-products (see Figure 5.1). Then, the uranium product is converted by fission into electrical energy, (in excess of that required for the mining/refining), with the accompanying unavoidable production of waste heat and waste radioactive fission products and the release of neutrons in excess of those required to sustain the chain reaction. Although only uranium is fed into the system, parasitic (non fission) capture of neutrons gives rise to a system working inventory which contains not only uranium, but also man-made transuranic isotopes – i.e., the macrosystem working inventory contains a full spectrum of natural and man-made actinide elements. Complete extraction of 200 MeV for every uranium atom fed into the system would imply that only fission products leave the system as an outflow. Incomplete extraction of 200 MeV for every uranium atom fed into the system would give rise to a buildup of actinide inventory inside the system. This is indeed the current world macrosystem situation (as illustrated in Figure 5.2); less than 200 MeV per U atom has been extracted up to now and no actinides have yet been returned to the earth's crust, but rather they remain in storage in working inventory. This situation gives rise to contemplation of how to proceed: either to consign the existing actinide inventories to the waste stream from a "once-through" macrosystem (see Figure 5.3), or recycle and extract further fission energy from the inventories of uranium and transuranics and send fewer actinides and more fission products to waste.

5.1.1 Multiple recycle

As discussed in Chapter 2, technology for recycle of plutonium in current-generation light water reactors has been well established by a 20 year development programme, and commercial MOX recycle has begun on several national power grids. When commercial MOX recycle in LWRs becomes widespread, its effect will be to retard the rate of net production of transuranic mass per unit of macrosystem energy benefit by recycling and fissioning a greater fraction of the transuranics produced upon in situ neutron capture on the U-238 in LWR fuel.

As shown in Chapters 2 and 4, multiple recycle of MOX in a thermal spectrum reactor is probably limited to two or three cycles, owing to the inexorable buildup of higher mass isotopes of plutonium and minor actinides. Such higher mass isotopes (particularly the even mass number threshold fissionable ones – Pu-240, Pu-242) act as absorbers in a thermal spectrum and their buildup upon multiple recycle leads to ever increasing enrichment requirements which leads among other effects to a positive coolant void coefficient if the enrichment reaches 10 to 12%. In the end, after two or three MOX recycles in an LWR one creates a macrosystem working inventory of transuranics containing a high fraction of higher mass number isotopes which are unsuitable for further thermal recycle; an inventory which therefore requires alternate means for ultimate disposition. The two choices are consignment to waste or further recycle in some other system whose physics favour fission over parasitic capture of the higher mass actinides.

From the viewpoint of the goal to extract the full energy content of the earth's endowment of uranium ore and to send only fission products to waste, such an existing actinide mass inventory would be viewed as an intermediate product. It would not yet have foregone any fraction of the energy potential of the original uranium ore because the fissions which would have occurred up to then are just a step along the way to full fission of every uranium atom. The U-238 and transuranics in such spent fuel would still be available for future fission consumption; the only thing which would have been discarded up to then are excess neutrons.

Fast spectrum reactors are well suited for still further extraction of energy from such transuranic inventories. As shown in Figure 5.4, all transuranics are fissionable in a fast neutron spectrum. Thus, in a fast reactor their further multiple recycle could achieve total fission destruction of the macrosystem's uranium feedstream – leaving only fission products for return to the earth's crust.

In the past, fast spectrum reactors have been designed to exploit their favourable neutron economy and to consume the uranium feedstream at a rate in excess of energy needs so as to build up an excess of transuranics in working inventory – in this way, they behave like LWRs. However, (in contrast to LWRs) the core geometries and compositions (i.e., the "loadings") of fast spectrum reactors can be adjusted so as to waste either more or fewer neutrons over a very broad range. At a given power rating if more neutrons are wasted – (diminishing the neutron captures on U-238), the recycle system would not be fissile self sufficient, and an *external feed of transuranics* would be required along with the U-238 feed in order to sustain the transuranic fissile working inventory of the fast reactor closed fuel cycle. Alternately, if at the fixed power rating, fewer neutrons are wasted and are instead captured on U-238 then the reactor system would become a breeder and generate excess transuranic mass, the system would become more than fissile self sufficient and transuranic mass would build up inside the system (perhaps as working inventory of additional reactor systems). The feedstream to the breeder would require no transuranics and would consume U-238 in excess of that needed for fission.

Using the ability to adjust the neutron utilisation as discussed above, the loadings of each fast reactor plant can at each recycle reloading be variously configured to produce transuranic conversion ratios lying anywhere between 0.5 and 1.5. Such a range of available transuranic conversion ratio provides a means for life-cycle management of the global transuranic inventory. As illustrated in Figure 5.5, it provides the means to extract the full energy potential from the earth's endowment of uranium ore; to hold constant (Conversion Ratio = 1) or to increase (Conversion Ratio > 1) the deployment of fast reactor closed fuel cycle plants on a regional basis in response to power demand; and eventually to decrease the number of power plants deployed and to close the nuclear power era out in an ecologically-sound manner by using burner (Conversion Ratio < 1) core loadings to consume the working inventories of decommissioned sibling units.

As regards to waste returned to the earth's crust in such a regime, in all scenarios and at all times, the transuranic content consigned to the waste stream would be maintained nearly zero (recycle losses only) so that the global transuranic inventory would be contained entirely in the working inventories of power-producing fast reactor closed cycle power plants and would eventually be totally transmuted to energy and fission products.

5.1.2 Thermal reactor/fast reactor symbiosis

The fast reactor closed fuel cycle *requires* that the initial transuranic working inventory be provided from an external source. It has long been envisioned that the fissionable material for this initial working inventory would come from inventories of transuranics built up from thermal reactors, after several thermal recycles make the transuranic composition unsuitable for further thermal recycle.

The transuranic material generated in the thermal reactor segment and supplied to the fast reactor segment of the symbiosis could be used in either of two ways:

- As initial working inventories for a growing number of fast reactors which operate in a fissile self sufficient or breeder mode;

- Or alternately as annual fuel feedstock for a fixed number of fast reactors which operate as net consumers of transuranics in the burner mode.

  i.e., a symbiosis of LWRs and fast burner reactors would more than simply retard the rate of net transuranic production per unit of energy benefit; the ratio of fast burners to LWRs would be adjusted so as to decrease the inventory of transuranics contained inside the macrosystem.

In the current world situation with inventories of excess transuranics amounting to a mass of some 900 000 kg of transuranics worldwide, the initial decades of a fast reactor/LWR symbiosis could likely employ fast reactors as burners in order to work off the excess. At a later stage, when both the excess plutonium and the U-235 reserves have been exhausted, the fast reactor loadings could be converted over for fissile self sufficient or breeder mode of operation. Then U-238 tails left over from enrichment operations and the remaining 96 w/o of unused uranium present in the reserves of LWR spent fuel represents an enormous fuel stockpile for future fast reactor energy generation.

Figure 5.6 illustrates the potential near-term symbiotic system in which the spent fuel from the thermal reactor enterprise is viewed as feedstock for a fast burner reactor cycle. As shown in the figure, an additional technological element is required for a symbiosis: a recycle technology is needed which can recover the actinide content of the LWR spent fuel and transform it into a suitable feedstream for the fast reactor closed fuel cycle. Two alternatives can be considered:

- The first is the *aqueous PUREX/TRUEX recycle technology* which is well established and is already commercially deployed in large capacity plants;

- The second is the *PYRØ recycle technology* for which scientific feasibility has been established and which is undergoing engineering development.

5.2 Fast burner reactor physics benchmarks

5.2.1 Recycle technology options

The overall structure of the Working Party's logic as it pertains to symbiosis of fast reactors with the current generation thermal reactor fuel cycle is illustrated in Figure 5.7. First, for the near-term, the fast reactors will be burners and will rely on an annual feedstream of transuranics and uranium from

117

LWRs and/or already existing stocks of excess transuranics. Given that transuranics (TRU) are to be extracted from LWR once, twice or trice through spent fuel, then if they are extracted by PUREX/TRUEX technologies wherein the Pu and minor actinides report to separate product streams, a choice exists whether to send all transuranics (TRU) to the fast reactor or to return the Pu to mixed oxide (MOX) recycle in LWRs and send only the minor actinides (MA), to the fast reactor. Alternately, if the TRU is extracted from LWR spent fuel by PYRØ technologies, only the first option is available.

Thus, three relevant fast reactor transuranic burnup options exist, several of which are the subject of the Working Party's Fast Reactor Burner Benchmark exercise:

- Oxide-fuelled fast reactor with fast reactor PUREX recycle and *minor actinide feed* from LWR PUREX/TRUEX processing;

- Oxide-fuelled fast reactor with fast reactor PUREX recycle and *transuranic feed* from LWR PUREX/TRUEX processing;

- Metal-fuelled fast reactor with fast reactor PYRØ recycle and *transuranic feed* from LWR PYRØ processing.

In all three cases, the fast reactor system comprises (as shown in Figure 5.8):

- A reactor of specified power rating, capacity factor, and conversion ratio;

- An external feedstream generated from the recovery of TRU or MA contained in LWR spent fuel (which otherwise would have itself been consigned to waste) plus any required U-238 feedstream also recovered from the spent LWR fuel;

- A recycle/refabrication facility for recycling fast reactor spent fuel. This recycle plant is of specified partitioning fractions and dwell times wherein the fast reactor discharge fuel is processed to separate fission products from the actinides and wherein the recovered actinides (i.e., TRU plus uranium) are blended with the feedstream from the thermal reactor cycle to make fresh assemblies for return to the fast reactor;

- A waste materials stream leaving the fast reactor system for ultimate disposal. This contains fission products plus any unrecoverable actinides;

- A waste heat stream dissipated to the environment, and;

- An electrical power output to the grid.

The fast burner reactor cycle is viewed as being interposed between the LWR system and the geologic Repository with a goal to fission a greater fraction of the uranium ore originally mined and now residing in LWR spent fuel in the form of unused uranium and transuranic isotopes. To the degree that multiple recycle in the fast cycle can transmute the uranium and transuranics which leave the thermal cycle into energy and fission products, it serves to diminish the amount of actinides sent to the Repository per unit of energy benefit from the macrosystem as a whole.

118

5.2.2 Issues, goals and groundrules for the burner benchmarks

Once the thermal rating (or electrical rating) of the fast reactor power plant (600 MWe in the case of these benchmarks) and the plant capacity factor (0.80 or 0.85 in the case of these benchmarks) are specified, then the number of actinide atoms consumed by fission per year is totally specified, i.e., the amount of energy benefit (kWe hours/year) is totally specified and the "issues" reduce to:

- What is the rate of reduction of the TRU or MA mass inventory from LWR spent fuel inventories – which would otherwise have been destined for ultimate disposal in the Repository – expressed in units of:

  - mass,
  - curies,
  - watts,
  - radiotoxicity hazard?

- What is the rate of buildup from the fast reactor fuel cycle of the waste stream which is now destined for ultimate disposal in the Repository from the fast reactor fuel cycle – expressed in the same units as above?

- What is the net rate per unit of energy benefit from the symbiosis cycle – (output from fast cycle minus input from thermal reactor)/energy from fast cycle – expressed in the same units?

And, finally:

- What are the safety and operating characteristics of the fast *burner* reactor power plant in contrast to the standard breeder designs – i.e., how are the fast reactor's safety properties impacted by the TRU or MA burning mission?

Given that the energy extraction per year (and, therefore, the number of actinide atoms consumed by fission per year in the fast reactor) is already totally specified, the first issue is directly and exclusively dependent on the conversion ratio to which the fast burner reactor is designed. The lower the conversion ratio below one, the smaller will be the amount of fuel recycle from the closed fast cycle, and the greater will be the rate of reduction of the LWR spent fuel transuranic inventory, for all units used to express it.

For the second issue, the masses of fission products sent to waste are fully specified by the energy benefit. The *masses* of actinides sent to waste depend on the product of working inventory times recycle fraction per annum times (1 - recovery fraction) of the chemical recycle process. The *isotopic distribution* of the actinide mass sent to waste depends strongly on conversion ratio (the lower the conversion ratio, the larger the fraction of LWR feed material vis-à-vis fast reactor recycle material is present in the discharged fuel for processing). Moreover, because of threshold fission effects, it depends strongly on neutron spectrum, and finally, it depends in subtle ways on neutron cross sections, interaction branching ratios, decay constants, average discharge burnup, and out-of-reactor dwell time (both before and after processing).

The third class of issue is dependent not only on fast reactor conversion ratio but also on neutron spectrum in the reactor core. The neutron spectrum is dependent strongly on both composition and leakage/absorption ratio, so it is controlled by the choice of MOX vs. metal fuel, the choice of means to dispose of excess neutrons – leakage vs. in-core parasitic absorption, and the blending ratio of external feed to recycle feed in fresh fuel fabrication – i.e., on conversion ratio.

The fast burner reactor benchmark exercise has as its two goals to:

- investigate the impact upon the four issues enumerated above imparted by the selection which is made among the three fast reactor system deployments shown in Figure 5.7 – and to do it parametrically in core conversion ratio, and to;

- ascertain the degree of spread in technical predictions which is achieved among participating organisations – each using its own set of cross sections and computational tools.

As regards to groundrules, the Working Party intends that the reactor designs used for benchmarking should be legitimate designs which meet all relevant design requirements and constraints i.e., the parameter studies on conversion ratio should employ reoptimisation and not be simply a partial derivative off a previously-optimised design.

All recycle/refabrication facility representations should use partition fractions between recycle and waste streams and dwell times which are realistic, best current knowledge values. Since the integrated toxicity of the materials sent to the ultimate disposal depend directly on the recycle partition fractions, a key conclusion of the study depends critically on the realism of the partition fractions used in the benchmark.

The "figure of merit" mass and toxicity flow results in Issues 1 and 2 should be normalised to a measure of benefit such as electrical energy benefit received which otherwise would have been foregone; e.g.,

- (LWR TRU Mass Reduction)/MWe year of fast reactor-generated energy,

- (Net Ci to Waste)/MWe year of fast reactor-generated energy benefit,

- others.

And, the studies should address both the asymptotic state (after multiple recycles in the fast cycle) and the startup transient prior to achieving the asymptotic state.

5.2.3 The physics design of fast burner reactor cores

The physics of a Pu-burner fast reactor core is defined by the relations which characterise the neutron balance in a fast reactor core and the related variations of the main reactivity coefficients (Na-void reactivity effect, Doppler coefficient, reactivity loss per cycle). Since the Pu-burning function is directly related to the increase of the Pu content in the fuel (and the corresponding reduction of the U-238 content), the internal breeding gain (IBG) will become more negative (i.e., the conversion ratio will become less than 1). Theoretically, plutonium without uranium fuel types can reach the maximum Pu consumption (~ 110 kg/TWhe). However, an increasingly negative IBG will cause the reactivity loss per cycle (Δk/cycle) to become larger (i.e., it will cause shorter irradiation cycles). This has on the other hand beneficial effects, i.e., decrease, on the positive Na-void reactivity effect (Δk_{Na}). Finally a lower U-238 content means in principle a smaller Doppler coefficient (Δk_D). Studies performed in the frame of the IFR programme in the U.S.A. and of the CAPRA project in France have demonstrated that the cautious resorting to poisoning, in which the excess reactivity linked to the increase of the fuel Pu content is compensated by the introduction of a neutron absorber, is a valid option to reduce the penalties related to short in-pile fuel residence time. Since neutron spectrum hardening through the introduction of absorbers deteriorates both Δk_{Na} and Δk_D, neutron spectrum softening by introducing moderating materials can help together with heterogeneous core arrangements to restore acceptable values of the $\Delta k_{Na}/\Delta k_D$ ratio, and to keep reasonable values for Δk/cycle.

In summary, the physics of Pu-burner fast reactor cores is determined by a delicate balance of parameters in cores where leakage (or reflector) properties play an increasingly important role.

5.2.4 Burner benchmark specifications

Two benchmark fast burner reactor systems were specified; the first based on oxide fuel and the PUREX recycle process; the second based on metallic alloy fuel and the PYRØ recycle process. Both are for burner core designs and both use similar strategies to lower the conversion ratio well below unity: the uranium content in the reactor is reduced both by removing blanket assemblies and by increasing the enrichment of the driver fuel up to the limits of the fuel's irradiation data base. The neutrons which otherwise would have been captured on uranium are purposely wasted by dramatically increasing the core leakage fraction.

The two benchmarks span the range of possibilities for fast burner reactor recycle in an intermediate time interval prior to widespread commercialisation of fast fissile self sufficient or breeder reactor designs. In the case of the oxide benchmark, the feedstream from the thermal reactor cycle is strongly skewed toward heavier plutonium isotopes; as shown in Table 5.1, the Pu-239 and Pu-240 are each present at about 40% and the Pu-241 and Pu-242 content is high. Moreover, the Np-237 and Am-241 content is also high. This feedstream is characteristic of a scenario in which the plutonium has been twice recycled (three times burned) in a thermal spectrum LWR, and in which during the thermal cycle PUREX reprocessing step, the Np and Am had been set aside, not recycled to the LWR, but instead saved for the fast reactor feedstream.

	Pu-238	Pu-239	Pu-240	Pu-241	Pu-242	Am-241
at%	5,6	39,1	26,7	13,0	14,3	1,3

Table 5.1 Oxide cores: plutonium isotopic composition*

* Pu/U + Pu ratio for inner and outer cores: 28, 28%, 40, 64% (mass).

In the case of the metal-fuelled benchmarks, the feedstream from the thermal reactor cycle represents LWR once through fuel with about three years of cooling prior to injection into the fast reactor closed fuel cycle. A pyrometallurgical technology to reduce LWR spent fuel and produce a fast reactor metallic feedstream containing all transuranics admixed together (Pu + Np + Am + Cm) has been assumed. The plutonium isotopic vector and indeed the whole transuranic vector with the exception of the presence of Cm is skewed more to the lighter isotopes than is the case for the feedstream to the oxide benchmark.

The neutron balances of these fast burner core designs are quite different from the conventional fissile self-sufficient or breeder designs to which the neutron interaction cross-section data sets and the calculational methods have been extensively verified as a part of the historical fast breeder reactor development program. It was anticipated that the variability in technical predictions would exceed that which has been found in previous NEACRP sponsored benchmarks on traditional breeder designs. It is clear that transport corrections will be substantially larger for these high leakage cores, and that threshold fission in even-mass number isotopes could be influenced by differing slowing down treatments in the cross section generation and leakage in the full core calculations (similar considerations were found in the recent IAEA benchmark on low sodium void worth cores). One of the goals of the benchmark activity, as stated above, is to assess the variability among participants' solutions which arises for burner cores whose neutron balance is substantially altered from that of traditional designs.

The oxide-fuelled benchmark

The oxide benchmark is a 600 MWe (1500 MWth) burner reactor which operates on a 125 EFPD cycle at 80% capacity factor; one fifth of the core is refuelled per cycle. As shown in Figure 5.9, the core is of a homogeneous layout with two radial enrichment zones and no radial blankets. Axially, the core is about a meter high and has no axial blankets. The conversion ratio is near 0.5. *The beginning-of-life compositions are specified* as shown in Table 5.2. The compositions represent discharge from LWR after two MOX recycles. The oxide-fuelled benchmark is specified in detail in Volume 4 of the Working Party Report.

The requested edits include the beginning of life eigenvalue and neutron balance, spectral indices and safety coefficients. Also requested are the composition and eigenvalue changes after a single burn cycle. Decay heat and isotopic contributions to toxicity (using specified toxicity factors) are also requested.

The metal-fuelled benchmark

The metal-fuelled burner benchmarks are based on a 600 MWe (1575 MWth) configuration. The benchmark specifications are described in detail in Volume 4 of the Working Party Report. As shown in Figure 5.10, the core region contains driver assemblies surrounding a central reflector/absorber island. The active core height is only 45 cm, roughly half the height of conventional fast reactor designs. Axially, the model has no axial blankets and has a 15 cm reflector region below the core followed by a shield region. The fuel pins extend above the active core region with a 70 cm upper fission gas plenum. Three rows of radial shielding surround the active core, a single row of steel and two rows of absorber. This pancaked, annular geometry greatly enhances neutron leakage giving a low conversion ratio of roughly 0.5.

122

The cycle length is one year with an 85% capacity factor. In addition to mass flow characteristics, the toxicity of the fuel cycle inventories, feedstream, and discharged waste stream are evaluated. The oral ingestion cancer dose measure developed by Cohen is used to quantify these radiotoxicities. Shown in Table 5.3 are the specified toxicity data for converting mass to toxicity for heavy metal and most important fission product isotopes.

Three metal-fuelled benchmark cases were specified – starting with a simple physics benchmark and progressing to benchmarks of more complexity and relevance to the issues identified above.

The first benchmark is a metal-fuelled <u>BOL</u> <u>Core</u> where the *Beginning of Life (BOL) loading is specified*, and is irradiated for a single cycle. In this sense, it is similar to the oxide benchmark. A detailed reporting of the BOL neutron balance, BOL eigenvalue, neutron flux energy spectrum, and one-group collapsed cross sections for the transuranic isotopes was requested of participants. Requested depletion reporting include the eigenvalue change and mass increments for a single depletion step. Particular focus is placed on the detailed isotopic mass flow of the transuranic isotopes and the resulting radiotoxicity flows into and out of the reactor. The radiotoxicities associated with these mass flows are evaluated at decay times of 1, 10 10^2, 10^3, 10^4, 10^5, and 10^6 years subsequent to discharge.

The second benchmark evaluates the performance of the burner core design for a <u>Once-Through</u> fuel cycle. Fresh assemblies are fueled with a mixture of depleted uranium and transuranics from LWR spent fuel; and *each participant determines the enrichment (LWR TRU /Total Heavy Metal)* requirement for an equilibrium three-batch cycle (equal parts blend of fresh, once-burned, and twice-burned material). Mass flow characteristics are evaluated in a manner similar to the first benchmark case. In addition, selected safety parameters are reported: the delayed neutron fraction, fuel Doppler coefficient, sodium void worth, burnup reactivity swing, and decay heat level. The decay heat is evaluated at decay times of 1 hr, 1 mo, 1, 10, 10^2, 10^3, and 10^4 years.

The third benchmark evaluates the <u>Multiple Recycle</u> performance of alternative core designs with a *parametric variation in conversion ratio*. Fresh assemblies are fuelled with a *participant-determined blend of fast reactor recycle material and makeup transuranic feed* coming from LWR spent fuel. Specified ex-core processing intervals and recovery factors to be used are indicated in Table 5.4. In the fuel refabrication, transuranics recovered from the fast reactor recycle are fully consumed in the blending, and the LWR Once-Through transuranics are used only as required for make-up. Roughly 5% of the rare-earth fission products are recycled in the fast reactor PYRØ fuel cycle. Mass flow characteristics are evaluated in a manner similar to the first two benchmarks with a focus on the impact of conversion ratio on the rate of drawdown of the toxicity represented by the thermal LWR transuranic inventory. The safety parameters are evaluated with a focus on their changes as a function of conversion ratio.

Detailed reporting of all participant's results and analyses of the trends in performance and in variability among participants for the oxide and metal fuelled benchmarks are provided in Volumes 4 and 5, respectively, of the Working Party Report. Salient features and conclusions are summarised below.

5.3 Assessment and conclusions from BOL benchmarks

Six solutions were submitted for the oxide-fuelled BOL benchmark as shown in Table 5.5. Five solutions were submitted for the metal-fuelled BOL benchmark as shown in Table 5.6. Tables 5.7 and 5.8 show eigenvalue and neutron balance results for the oxide and metal benchmarks, respectively. Variability of several percent Δk in predicted BOL eigenvalue for cores of specified geometry and composition are observed, and no patterns of consistent clustering of participants solutions are discernible. Table 5.9 (oxide core) shows that the transport correction, which is of the order of half a percent Δk, is inconsistently predicted.

Tables 5.10 and 5.11 display the burnup reactivity loss for the oxide and metal cores, respectively; spreads of up to 1% $\Delta k/kk'$ out of a burnup swing of around 6% (metal), 8% (oxide) total reactivity loss are observed.

A general characterisation of the results displayed in detail in Volumes 4 and 5 is that the calculational predictions of these high leakage, fast burner reactor core designs display unacceptably large variability among participants; much larger than that experienced in benchmarks on more traditional designs.

Curiously, radiotoxicity flows relevant to the deployment of fast burner reactors as a waste management measure display less variability among participants than do operating and safety parameters – so long as consistent curie-to-radiotoxicity conversion factors are employed. Tables 5.12 and 5.13 for oxide at discharge and at a million years cooling and Table 5.14 for metal at discharge illustrate this result.

The causes for the variabilities among participants' solutions are analysed in some detail in Volumes 4 and 5. It is clear that for high leakage cores, two- and three-dimensional transport codes should be used – and are now available. These should tend to reduce the broad variabilities in core leakage probability (see Tables 5.7 and 5.8). However, no broad characterisation of a principal cause of the large variabilities has been found; a followon effort making use of transport solutions and of sensitivity coefficients of reactor parameters to cross section library values would be useful.

Based on the large variabilities among participants predictions of core performance, it is concluded that design and deployment of high leakage fast burner reactors will likely require supporting critical facility measurements to lower uncertainties in core operating and safety performance predictions. Alternately, the relatively less variability in predicted toxicity flows suggests that the second goal of the benchmark exercise can already lead to useful characterisation of trends for fast reactor/thermal reactor symbiosis as a waste management measure.

5.4 Assessment and conclusions regarding radiotoxicity reductions in the net waste stream from a fast burner reactor/LWR symbiosis

Two participants (PNC and US) provided solutions for the Once-Through metal-fuelled benchmark, and one participant (US) provided results for the multiple recycle metal-fuelled benchmark. Detailed analyses of the performance results for these benchmarks are given in Volume 5. The discussion here focuses on differences in safety parameters and differences in net toxicity flows caused by the parametric variation of the conversion ratio.

The recycle metal-fuelled benchmark evaluated net toxicity flows when a fast burner reactor is used to consume transuranics from the LWR cycle. Three 600 MWe reactors were considered – having conversion ratios near 0.5, 0.75, and 1.0. An external feedstream of transuranics from once through LWRs was used for makeup to the closed fast burner reactor cycle wherein the transuranics were multiply recycled to eventual total fission destruction. In all cases, the uranium recovered during fast recycle is set aside for future use, the recovered transuranics are blended with the LWR TRU feedstream plus makeup U-238 and returned to the reactor, and the fission products plus the small amount of TRU loss from the recycle chemistry is sent to the Repository. (In the case of the CR=1 core, the external feedstream is depleted U only.) Transuranics leave the fast reactor cycle and enter the waste stream only to the extent of carryover to waste in the PYRØ recycle step. The *net* waste flow to the Repository is comprised of the difference between the waste stream from the fast cycle and the feedstream entering the fast cycle from the LWR – which otherwise would have gone to the Repository.

The recycle segment of the fast reactor fuel cycle was specified to include a one year cooling interval followed by a six month chemical separation period, followed by a six month refabrication time; the total ex-core interval between discharge and re-insertion is two years (see Table 5.4). The pyroprocessing treatment looses 0.1% of the transuranics to the waste but returns 99.9% to the core for further fission consumption; 5% of the rare-earth fission products are also recycled; all other fission products go to the waste stream. Uranium is separated for interim storage until it can ultimately serve as feedstock for the fast breeder reactor cycle after the symbiosis with thermal reactors ceases owing to unavailability of U-235 for further LWR deployment.

The neutronic performance characteristics of the three recycle cases of differing conversion ratio are intercompared along with the US solution of the Once-Through benchmark in Table 5.15.

As expected, the CR = 0.5 core consumes the external transuranic feedstock at the quickest rate (255 kg/y); the consumption rate of the CR = 0.75 design is roughly half that rate, and the breeder design has a net transuranic production of 10 kg/y. Note that at the 600 MWe power rating the maximum feasible consumption rate (at roughly 1 g/MWth day) is 490 kg/y. This rate would require a reactor design completely devoid of U-238 in order to preclude in situ production of transuranics.

The transuranic consumption is replenished by the external feed of LWR transuranics. None-the-less the majority of the mass annually loaded into the fast reactor comes from recycle of fast reactor spent fuel material since the average discharge burnup is only ~10% of the initial heavy metal, and the rest is available for recycle. For example, in the CR = 0.5 design, the burnup of transuranic isotopes is slightly greater than 10%, but still 86% of the transuranic reload material comes from the recycled inventory. As shown in Table 5.15, the total heavy metal mass flow rate through the recycle equipment increases significantly for the CR = 0.75 and CR = 1 configurations because fertile material from external zones must be recycled also.

The burner (CR = 0.5) configuration requires a higher enrichment (increase from 26 to 30% TRU/HM) as compared to the Once-Through case; this is attributed to a buildup of Pu-240 as a result of multiple recycle. On the other hand, the burnup swing is significantly lower (4.3% Δk as compared to 5.3% Δk) in the recycle case, because Pu-240 is a superior fertile material. The enrichment requirement steadily decreases to 22.5% as the conversion ratio is increased, and the reduced enrichment requirement of the CR = 0.75 and CR = 1.0 designs improves the internal conversion ratio, and reduces the burnup reactivity swing from 4.3% Δk in the CR = 0.5 recycle case to roughly 3.5% Δk.

Variations in the safety parameters result from the parametric variation of conversion ratio. The Doppler coefficient increases in magnitude from -1.35×10^{-3} Tdk/dT in the CR = 0.5 design to -1.80×10^{-3} Tkd/dT in the CR = 1.0 design; this change is attributed to the higher U-238 concentration in the reduced enrichment fuel and blanket zones. In a similar manner, the lower enrichment requirement for the Once-Through burner design leads to a significantly higher Doppler coefficient. A significant increase in the sodium void worth (over 0.5%Δk/kk') is observed between the Once-Through and recycle results; this change is caused by the buildup of the threshold fissionable Pu-240 upon multiple recycle. The sodium void worth values given in Table 5.15 are for the driver regions alone; the blankets are assumed to remain flooded. Thus, the void worth is slightly reduced in the CR = 0.75 and CR = 1.0 design as the blanket power fraction increases. Isotopic changes between the designs lead to but small variations in the delayed neutron fraction.

The inventories and mass flows for the recycle benchmark were converted and expressed in toxicity units using the conversion data of Table 5.3. The toxicity inventories and flow rates at several points in the fuel cycle are summarised in Table 5.16. [1] The CR = 0.5 recycle case can be used to illustrate the interpretation of the toxicity flow data.

Within the reactor, the BOEC inventory has a total toxicity of 225E7, and the EOEC inventory has a total toxicity of 230E7. The toxicity of the fission products component increases significantly during the burn cycle as more fission products (FP) are formed. Alternately, the TRU toxicity of the in-reactor inventory is relatively constant due to the competing effects of total inventory reduction of TRU (burning) and a temporary TRU isotopic mass fraction increase of short-lived highly toxic isotopes (e.g., Cm-244).

Roughly one-third of the TRU inventory, and one-half of the FP inventory are discharged from the reactor each year for recycle processing. As shown in Table 5.16, however the vast majority (99.9%) of the TRU toxicity, 63.2E7, is recycled back to the reactor; and only 6.32E5 of it is discharged to the waste as a result of the 0.1% TRU loss during PYRØ chemical operations. Thus, the short-lived FP component (1.48E7) comprises 96% of the discharge waste toxicity (1.54E7).

The feedstream comprised of LWR transuranics and uranium (which otherwise would have been sent to waste) has a toxicity of 3.51E7 and is totally dominated by the toxicity of its TRU component. The LWR/fast burner reactor symbiosis thus leads to a net reduction of toxicity flow into the Repository by about a factor of two (from 3.51E7 to 1.54E7) as a result of transmuting the transuranics from the LWR spent fuel into energy and fission products. *The fast burner reactor fed by TRU from the LWR cycle produces a negative net flow of toxicity to the Repository per unit of energy benefit.* While this short term payoff is useful, the major payoff occurs on the longer time scale.

The feedstream toxicity is dominated by the TRU component most of which are very long-lived isotopes, whereas the waste stream toxicity is dominated by the short-lived FP component. As a result, the size of the net toxicity balance – toxicity of the fast reactor waste stream as compared to the feedstream comprised of LWR transuranics which otherwise would have been sent to waste – depends significantly on when the comparison is made because it is altered by radioactive decay over time. The time-dependent toxicity of the fast reactor waste vis-à-vis its feedstream for the CR = 0.5 case is compared in Figure 5.11 for decay times up to a million years; Figure 5.12 shows the components comprising the total toxicity of Figure 5.11. As shown in these figures, the toxicity of the fast reactor waste stream decays by three orders of magnitude over the first 500 years due mostly to short-lived

---

[1] Note that these toxicity values only included the isotopes for which toxicity data are listed in Table 5.3.

fission product die away. In contrast, over the same time period the toxicity of the TRU-dominated feedstream from LWR spent fuel would have decayed by less than one order of magnitude. So, when viewed at 500 years, the net result of LWR/fast burner reactor symbiosis as compared to burial of LWR spent fuel is a *two orders of magnitude reduction* in long-term toxicity – much larger than the factor of two reduction at discharge.

Figures 5.13 and 5.14 display the components of the toxicity flows for the CR = 0.75 and 1.0 cases, respectively. Just as for the CR = 0.5 case, the CR = 0.75 burner, achieves a net reduction of toxicity flow at all times subsequent to discharge. That is not the case for the CR = 1.0 fissile self sufficient case because of the absence of TRU in its feedstream. None-the-less, if one were to take credit for using LWR transuranics to create the initial working inventory for the fissile self sufficient fast reactor rather than sending them to the waste Repository, then a benefit is, of course, realised immediately.

Figure 5.12 shows that the toxicity attributable to the short-lived fission products falls below that of the long-lived fission products within 500 years. Moreover, it shows that in the very long-term the outflow of toxicity attributable to long-lived fission products is actually less than the inflow due to uranium. *This means that in the overall nuclear macrosystem, if no transuranics were lost to the waste, then subsequent to a 500 years waste sequestration period of short-lived fission products a negative net flow of toxicity to the earth's crust would result from deriving energy from nuclear power.* Figures 5.12, 5.13, and 5.14 show that a factor of 15 smaller TRU loss rate to the waste stream (less than the 0.1% used in this study) would be sufficient to achieve net negative toxicity flow subsequent to 500 years not only for the burners but for the fissile self-sufficient design also. Although such an improvement is known to be chemically achievable, further testing will be required to see if it can be accomplished in a practical engineering sense at the commercial scale.

In summary, the toxicity inventory and flow data from the fast burner reactor benchmark reveals several features:

1. For an LWR once-through or several MOX recycle fuel cycle, the toxicity in the waste stream is dominated by the TRU component at discharge and this component maintains its toxicity for many thousands of years. In contrast, by interposing a fast burner fuel cycle between the LWR and the Repository and multiply recycling the transuranics back to the reactor, their contribution to waste toxicity is reduced by two orders of magnitude, the short-lived FP dominate the fast reactor waste stream toxicity and these short-lived FP toxicity decay away in several hundred years.

2. Thus, a symbiosis wherein LWR discharge supplies feedstock to fast burner reactors yields a toxicity outflow from the fast reactor cycle to the waste stream which is significantly lower than the inflow of toxicity from the LWR (which otherwise would be destined for waste). For the CR = 0.5 the toxicity is reduced by roughly one-half the introduction rate each year, and for the CR = 0.75 design, the toxicity is reduced by roughly one-fourth each year.

3. Moreover, after 500 years, the short-lived fission products from the fast burner cycle decay away and because of the absence of significant transuranic mass in the waste stream, the toxicity reduction vis-à-vis disposal of the LWR fuel is not just a factor of two, but rather is two orders of magnitude.

4. A factor of 15 reduction in the recycle loss rate (of TRU) to the waste stream [2] from fast recycle would be sufficient to enable the fast reactor/LWR symbiosis to produce a net negative flow of long-term (> 500y) toxicity to the earth's crust as the overall legacy of nuclear power – i.e., more toxicity removed in the ore than is returned in the long-lived radioisotopes contained in the waste.

5. The results obtained in the oxide-fuelled benchmark and in the first case of the metal-fuelled benchmark show similar trends as far as radiotoxicity inventory at the end of a single burner cycle is concerned. This fact indicates that with the same hypothesis of high efficiency in the transuranic recovery at reprocessing (e.g., 0.1% loss to the waste stream), most of the conclusions derived from the metal-fuelled multiple recycle benchmark (including the effects of the parametric conversion ratio variation) are applicable to oxide-fuelled fast burner cores as well.

---

[2] vis-à-vis the value of 0.1% used in these benchmarks.

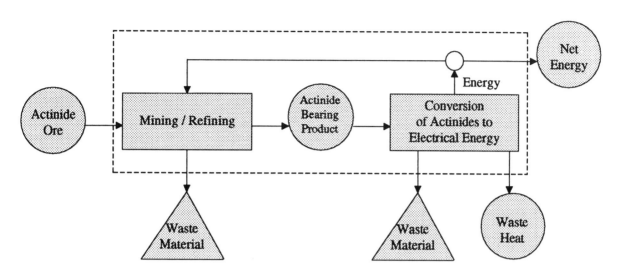

Figure 5.1 Macro system view of fission-induced energy extraction from earth's endowment of actinides

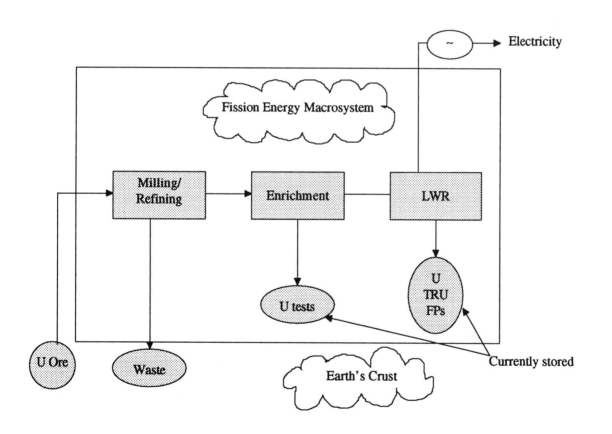

Figure 5.2 Current situation of the macrosystem

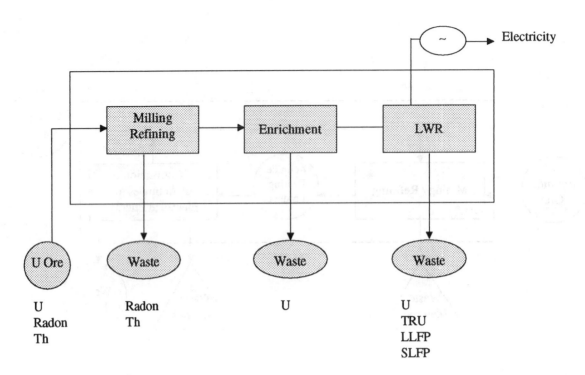

Figure 5.3 ***The once-through macrosystem***

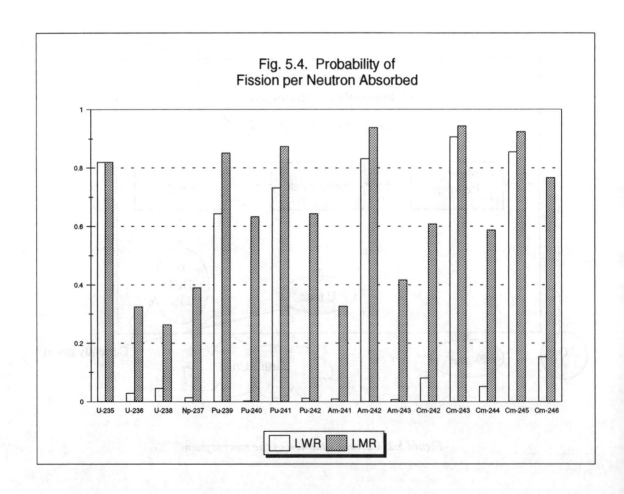

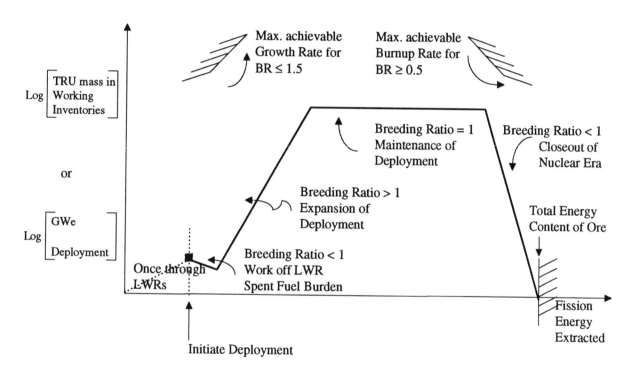

Figure 5.5 *Lifecycle management of transuranic burden*

131

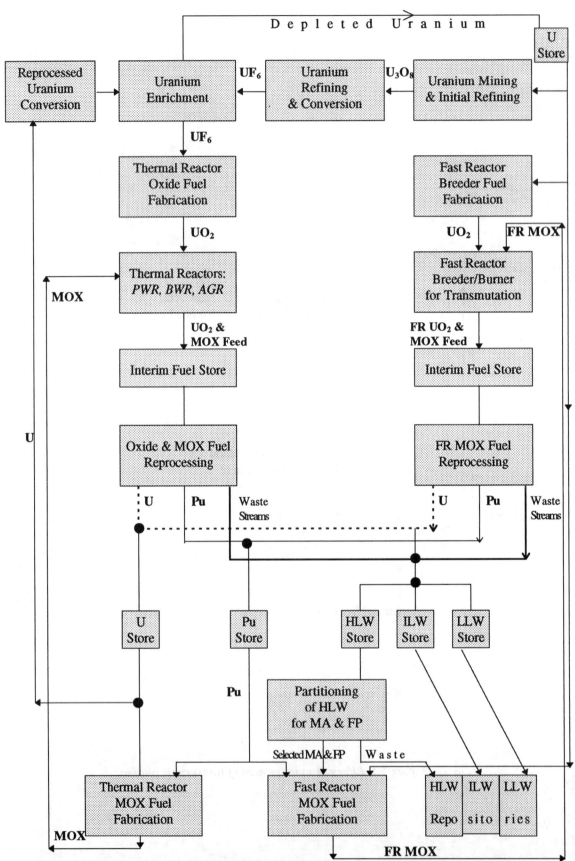

*Figure 5.6 **Thermal reactor fuel cycle with recycling and actinide burning in fast reactors***

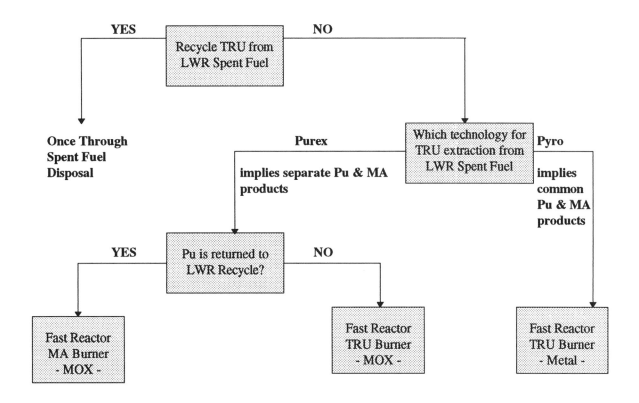

Figure 5.7 LWR/fast burner symbiosis options

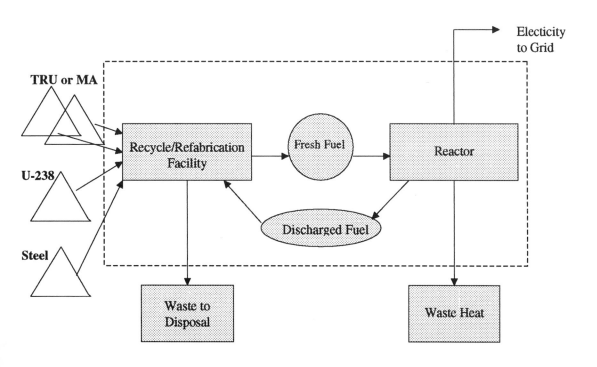

Figure 5.8 LMR multiple recycle burner core benchmark

RZ geometry

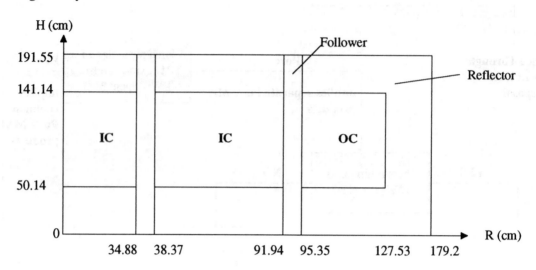

Figure 5.9 Oxide core geometry

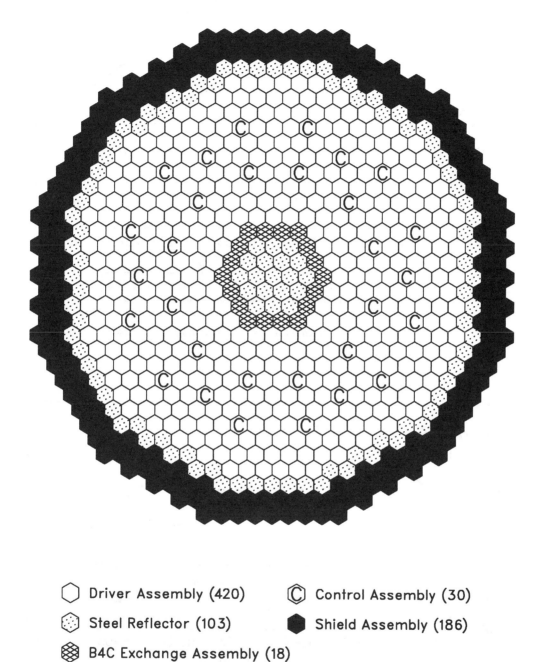

Figure 5.10 **Benchmark reference core configuration**

○ Driver Assembly (420)   Ⓒ Control Assembly (30)

▣ Steel Reflector (103)   ⬢ Shield Assembly (186)

▩ B4C Exchange Assembly (18)

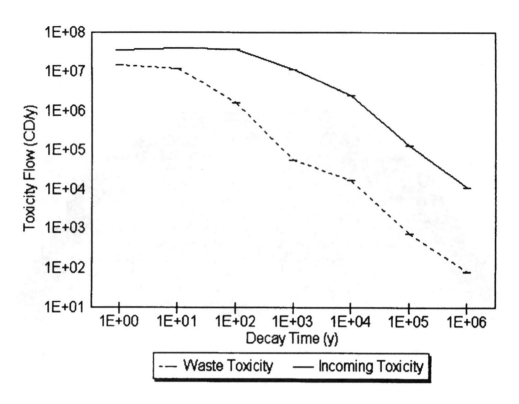

Figure 5.11 Toxicity flows for CR = 0.5 recycle case

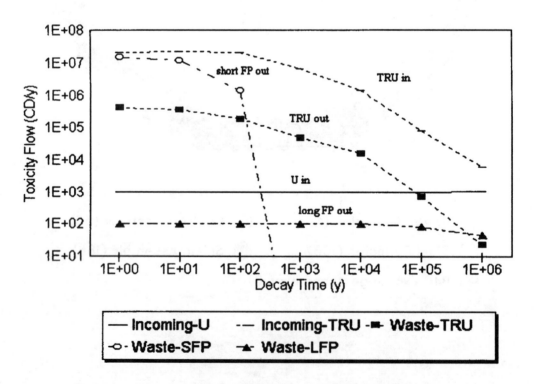

Figure 5.12 Toxicity components for CR = 0.5 recycle case

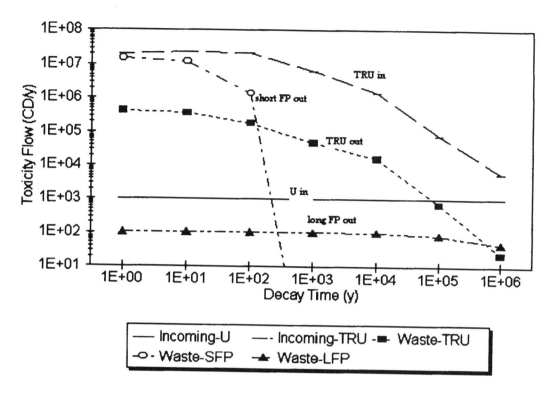

Figure 5.13 Toxicity components for CR = 0.75 recycle case

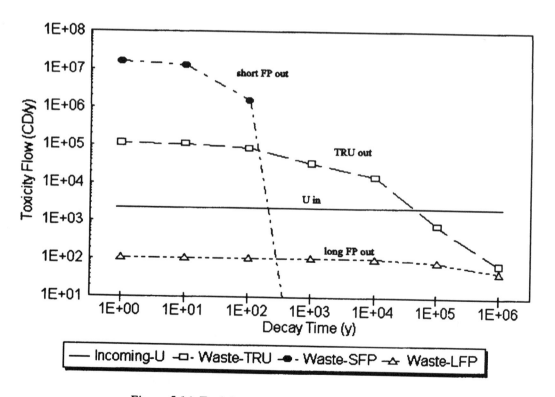

Figure 5.14 Toxicity components for CR = 1.0 recycle case

Region	Nuclide	Cell calculation inner zone	Cell calculation outer zone	Homogenised atomic density
INNER CORE	U-235	3.268-E-05		9.409E-06
	Pu-238	3.015E-04		8.683E-05
	U-238	1.304E-02		3.754E-03
	Pu-239	2.097E-03		6.037E-04
	Pu-240	1.426E-03		4.105E-04
	Pu-241	6.913E-04		1.990E-04
	Pu-242	7.573E-04		2.180E-04
	Am-241	6.913E-05		1.990E-05
	Fe		1.728E-02	1.231E-02
	Cr		4.973E-03	3.541E-03
	Ni		3.627E-03	2.583E-03
	Mo		4.360E-04	3.105E-04
	O	3.672E-02		1.057E-02
	Na		1.038E-02*	7.389E-03
	Mn		4.153E-04	2.957E-04
OUTER CORE	U-235	2.743E-05		7.899E-06
	Pu-238	4.247E-04		1.223E-04
	U-238	1.095E-02		3.152E-03
	Pu-239	2.953E-03		8.503E-04
	Pu-240	2.008E-03		5.782E-04
	Pu-241	9.736E-05		2.803E-04
	Pu-242	1.067E-03		3.071E-04
	Am-241	9.736E-05		2.803E-05
	Fe		1.728E-02	1.231E-02
	Cr		4.973E-03	3.541E-03
	Ni		3.627E-03	2.583E-03
	Mo		4.360E-04	3.105E-04
	O	3.684E-02		1.061E-02
	Na		1.038E-02*	7.389E-03
	Mn		4.153E-04	2.957E-04
AXIAL AND RADIAL SHIELDING	Fe			2.662E-02
	Cr			7.662E-03
	Ni			5.588E-03
	Mo			6.717E-04
	Na			1.093E-02
	Mn			6.398E-04
ROD FOLLOWER	Fe			7.987E-03
	Cr			2.299E-03
	Ni			1.676E-03
	Mo			2.015E-04
	Na			1.863E-02
	Mn			1.920E-04

* 0. for the voided cell

Table 5.2 **Oxide cores: atomic number densities**

Isotope	Toxicity Factor CD/Ci	Half-Life Years	Toxicity Factor CD/g
ACTINIDES AND THEIR DAUGHTERS			
Pb-210	455.0	22.3	3.48E4
Ra-223	15.6	0.03	7.99E5
Ra-226	36.3	1.60E3	3.59E1
Ac-227	1185.0	21.8	8.58E4
Th-229	127.3	7.3E3	2.72E1
Th-230	19.1	7.54E4	3.94E-1
Pa-231	372.0	3.28E4	1.76E-1
U-234	7.59	2.46E5	4.71E-2
U-235	7.23	7.04E8	1.56E-5
U-236	7.50	2.34E7	4.85E-4
U-238	6.97	4.47E9	2.34E-6
Np-237	197.2	2.14E6	1.39E-1
Pu-238	246.1	87.7	4.22E3
Pu-239	267.5	2.41E4	1.66E1
Pu-240	267.5	6.56E3	6.08E1
Pu-242	267.5	3.75E5	1.65E0
Am-241	272.9	433	9.36E2
Am-242m	267.5	141	2.80E4
Am-243	272.9	7.37E3	5.45E1
Cm-242	6.90	0.45	2.29E4
Cm-243	196.9	29.1	9.96E3
Cm-244	163.0	18.1	1.32E4
Cm-245	284.0	8.5E3	4.88E1
Cm-246	284.0	4.8E3	8.67E1
SHORT-LIVED FISSION PRODUCTS			
Sr-90	16.7	29.1	2.28E3
Y-90	0.60	7.3E-3	3.26E5
Cs-137	5.77	30.2	4.99E2
LONG-LIVED FISSION PRODUCTS			
Tc-99	0.17	2.13E5	2.28E-3
I-129	64.8	1.57E7	1.15E-2
Zr-93	0.095	1.5E6	2.44E-4
Cs-135	0.84	2.3E6	9.68E-4
C-14	0.20	5.73E3	8.92E-1
Ni-59	0.08	7.6E4	6.38E-3
Ni-63	0.03	100	1.70E0
Sn-126	1.70	1.0E5	4.83E-2

The toxicity factors are constructed using the methodology described by Bernard L. Cohn, "Effects of the ICRP Publication 30 and the 1980 BEIR Report on Hazard Assessments of High Level Waste", Health Physics, <u>42</u>, No.2, pp. 133-143, (1982) with the following data:

- ICRP Publication 30, Part 4, 19 <u>88</u>,
- BEIR III, 19 <u>80</u>.

The factors in the table stand for the fatal cancer doses per gram isotope injected orally. They denote the hazard of the material rather than the risk because they do not include any account of path way attenuation processes, but simply assume oral ingestion.

Table 5.3 **Radiotoxicity data**

REACTOR SEGMENT OF CYCLE

Cycle Length	365 days
Capacity Factor	85%
Power Rating	1575 MWth
Core Driver Refuelling	$^1/_3$ per cycle
Blanket Refuelling	$^1/_4$ per cycle

RECYCLE SEGMENT OF CYCLE

Cooling Interval	365 days
Chemical Separation	done on day 1 of second year
Blending & Fab.	done on day 184 of second year
Re-insertion into reactor	done on day 1 of third year

CHEMICAL PARTITIONING FACTORS	*% to Product*	*% to Waste*
All TRU isotopes	99.9%	0.1%
Rare Earth Fission Products* (excluding Y, Sm, and Eu)	5%	95%
All Other Fission Products*	0%	100%

* Recommend for Benchmark purposes, recycle zero fission products and send all to waste. ANL solutions are provided for recommended and for fission product recycle cases in the benchmark volume.

*Table 5.4 **Fuel cycle assumptions***

Country	Organisation	Contributors	Basic Data	Number of Energy Groups	Codes
France	CEA	J. C. Garnier F. Varaine	Carnaval-IV	25	HETAIRE, ERANOS
Japan	PNC	T. Ikegami T. Yamamoto S. Ohki	JENDL-2 JENDL-3.2	18	SLAROM, PERKY, TWOTRAN, ORIGEN-2
Japan	TOSHIBA	M. Kawashima M. Yamaoka	JENDL-3.1 JENDL-2 FSX(J3.1) VIM-LIB(J2)	70 70 Continuous Continuous	CRUX FASTMUG'Y MCNP-3B VIM
Russia	IPPE	A. M. Tsibulia	FOND-2 ABBN-90	26	SYNTES, MMK, CONSYST2, CARE
Switzerland	PSI	S. Pelloni	JEF2.2	30	NJOY, MICROR, MICROX-2, PSD, 2DTB, ORIHET
U.S.A.	ANL	G. Palmiotti	ENDF/B-V	28	MC^2-2, SDX, DIF3D, REBUS VARI3D, NUTS, TWODANT, ORIGEN-RA

Table 5.5 *Synopsis of contributions, basic data, and codes used in solving the oxide benchmark.*

Country	Organisation	Authors	Basic Data	Number of Energy Groups	Flux Solution
European	CEA-France AEA-UK	G. Rimpault, & J. DaSalva – *CEA* P. Smith – *AEA Technology*	JEF2.2	$1968 \rightarrow$ $33 \rightarrow 6$	HEX-Z Finite Diff.
Japan	PNC	S. Ohki & T. Yamamoto	JENDL-2	$70 \rightarrow 18$	RZ Finite Diff.
Japan	PNC	S. Ohki & T. Yamamoto	JENDL-3	$70 \rightarrow 18$	RZ Finite Diff
Russia	IPPE	A. Tsibulia	ABBN-90 (from FOND-2)	$26 \rightarrow \begin{cases} 17 \\ 6 \end{cases}$	HEX-Z Nodal
USA	Argonne	K. Grimm and R. Hill	ENDF-V	$2082 \rightarrow$ $230 \rightarrow \begin{cases} 21 \\ 9 \end{cases}$	HEX-Z Nodal

Table 5.6 *Participating countries and organisations for the BOL metal-fuelled benchmark*

Organisation	k-effective	Absorption (%)	Leakage (%)
ANL	1.10660	89.8	10.2
CEA	1.11170	89.0	11.0
PNC (J2)	1.12328	91.5	8.5
PNC (J3.2)	1.13106	91.0	9.0
Toshiba (J2)	1.11890	-	-
Toshiba (J3.1)	1.13488	-	-
PSI	1.12810	92.0	8.0
IPPE	1.11480	88.5	11.5

Table 5.7 **k-effective and critical balance at beginning of life: oxide**

	European	Japan JENDL-2	Japan JENDL-3	Russian	United States
BOL Eigenvalue	1.063[a]	1.098	1.092[b]	1.102	1.101
BOL Neutron Balance					
• Fissions per Core Absorption		0.589	0.592	0.600	0.599
• HM Captures per Core Absorption		0.376	0.374	0.370	0.366
• Structure Captures per Core Absorption		0.034	0.033	0.030	0.033
• Coolant Captures per Core Absorption		0.001	0.001	0.001	0.001
• Core Leakage per Core Absorption	0.623	0.585	0.601	0.616	0.612
• Model Leakage per Model Absorption		0.026	0.027	0.014	0.030
EOL Eigenvalue	1.012	1.040	1.034	1.040	1.042
Burnup Swing, %Δk	-5.1	-5.8	-5.8	-6.2	-5.9
Transuranic Inventory Ratio (EOL/BOL)	0.944	0.944	0.944	0.943	0.944

[a] Eigenvalue using nodal diffusion and nodal transport theory are 1.078 and 1.100, respectively.
[b] Estimated eigenvalue with mesh corrections applied is 1.085.

Table 5.8 **Metal core: comparison of reference core neutronic characteristics**

Organisation	k-effective	Sodium Void Whole Core	Doppler
ANL	0.531	0.117	-0.002
PNC (J2)	0.525	0.148	-0.007
PNC (J3.2)	0.526	0.139	-0.004
Toshiba (J2)	-	0.180	-
Toshiba (J3.2)	-	0.029	-
PSI	0.285	0.003	-0.002
IPPE	0.734	-	-

Table 5.9 Oxide core: transport effects at beginning of life in % of Δk/kk'

Organisation	BOL-EOC	BOL-EOC
ANL	7.61 (20.5%)	12.85 (20.1%)
CEA	7.90 (20.3%)	13.27
PNC (J2)	8.03 (23.1%)	13.60 (25.0%)
PNC (J3.2)	7.91(23.5%)	13.39 (25.2%)
PSI	7.79 (27.6%)	13.06 (28.7%)

Table 5.10 Oxide core: reactivity loss in % of Δk/kk' (in parenthesis fission products contribution)

	European	Japan JENDL-2	Japan JENDL-3	Russian	United States
EOL Eigenvalue	1.012	1.040	1.034	1.040	1.042
Burnup Swing, %Δk	-5.1	-5.8	-5.8	-6.2	-5.9
Transuranic Inventory Ratio (EOL/BOL)	0.944	0.944	0.944	0.943	0.944

Table 5.11 Metal core: burnup reactivity loss

Isotope	ANL	PNC	CEA
Am-241	7.69 E + 7	7.75 E + 7	7.48 E + 7
Am-242m	8.58 E + 6	8.65 E + 6	7.79 E + 6
Am-243	2.75 E + 6	2.70 E + 6	3.83 E + 6
Cm-242	1.62 E + 8	1.61 E + 6	2.07 E + 8
Cm-243	1.95 E + 6	1.74 E + 6	4.50 E + 6
Cm-244	7.33 E + 7	5.15 E + 4	1.33 E + 8
Cm-245	1.69 E + 4	1.03 E + 4	2.29 E + 4
Pu-238	2.44 E + 8	2.44 E + 8	2.56 E + 8
Pu-239	8.80 E + 6	8.60 E + 6	8.59 E + 6
Pu-240	2.39 E + 7	2.39 E + 7	2.44 E + 7
Pu-241	7.48 E + 7	7.50 E + 7	7.32 E + 7
Pu-242	2.05 E + 5	1.99 E + 5	1.99 E + 5
Tc-99	3.55 E + 3	3.64 E + 3	-
I-129	2.35 E + 3	2.69 E + 3	-
Cs-135	2.36 E + 3	2.40 E + 3	-
Total	6.77 E + 8	6.55 E + 8	7.90 E + 8

Table 5.12 *Oxide core: radiotoxicities at cooling time 0*

Isotope	ANL	PNC	CEA	PSI
Ac-227	1.11 E + 3	1.10 E + 3	1.14 E +3	1.13 E + 3
Np-237	2.70 E + 4	2.68 E + 4	2.65 E + 4	2.91 E + 4
Pa-231	8.20 E + 2	8.15 E + 2	8.43 E + 2	8.38 E + 2
Pb-210	1.17 E + 4	1.17 E + 4	1.28 E + 4	1.35 E + 4
Pu-242	3.26 E + 4	3.33 E + 4	-*	3.37 E + 4
Ra-226	2.63 E + 3	2.62 E + 3	2.88 E + 3	3.03 E + 3
Th-229	3.71 E + 4	2.82 E + 4	2.79 E + 4	3.06 E + 4
Th-230	1.24 E + 3	1.25 E + 4	1.36 E + 3	1.44 E + 3
U-233	1.95 E + 3	1.94 E + 3	-*	2.10 E + 3
U-234	4.24 E + 2	4.23 E + 2	-*	4.81 E + 2
U-235	1.93 E + 1	1.92 E + 1	-*	1.92 E + 1
U-236	3.81 E + 2	3.78 E + 2	-*	3.87 E + 2
Tc-99	1.37 E + 2	1.41 E + 2	-	8.37 E + 1
I-129	2.25 E + 3	2.57 E + 3	-	1.45 E + 3
Cs-135	1.74 E + 3	1.78 E + 3	-	1.11 E + 3
Total	1.21 E + 5	1.13 E + 5	-*	1.20 E + 5

* CEA took into account an extra 0.3% of U as a loss. This will affect the comparison for this isotope.

*Table 5.13 **Oxide core: radiotoxicities at cooling time one million years***

Isotope		Europe	Japan JENDL-2	Japan JENDL-3	Russia	United States	Mean Value
BOL Uranium							30.7
EOL Uranium		48.6	30.2	30.2	29.3	45.8	36.8±23%
Np-237	BOL						3.32E4
	EOL	2.99E4	2.99E4	2.97E4	2.98E4	2.99E4	2.98E4±0.3%
Pu-238	BOL						1.89E8
	EOL	2.35E8	2.51E8	2.52E8	2.45E8	2.47E8	2.46E8±2%
Pu-239	BOL						3.76E7
	EOL	3.52E7	3.52E7	3.52E7	3.48E7	3.53E7	3.51E7±0.5%
Pu-240	BOL						5.42E7
	EOL	5.37E7	5.39E7	5.37E7	5.32E7	5.35E7	5.36E7±0.4%
Pu-242	BOL						1.84E5
	EOL	1.96E5	1.93E5	1.91E5	1.88E5	1.91E5	1.92E5±1%
Am-241	BOL						1.06E8
	EOL	1.15E8	1.11E8	1.13E8	1.15E8	1.13E8	1.13E8±1%
Am-242m	BOL						1.40E6
	EOL	4.51E6	7.23E6	6.41E6	5.44E6	6.22E6	5.96E6±15%
Am-243	BOL						6.16E6
	EOL	5.72E6	5.72E6	5.78E6	5.89E6	5.83E6	5.79E6±1%
Cm-242	BOL						9.14E5
	EOL	1.89E8	1.04E8	9.00E7	8.73E7	8.73E7	1.12E8±35%
Cm-244	BOL						3.33E8
	EOL	4.13E8	4.12E8	4.08E8	3.54E8	3.79E8	3.93E8±6%
BOL Transuranics							7.32E8
EOL Transuranics		1.06E9	9.84E8	9.64E8	9.00E8	9.31E8	9.68E8±6%
Pu-241 + Am-241		5.79E8	5.78E8	5.81E5	5.77E8	5.83E8	5.80E8±0.4%
Pu-239 + Pu-240		8.89E7	8.91E7	8.89E7	8.80E7	8.88E7	8.87E7±0.4%
Pu-241 + Am-241 + Np-237		1.16E5	1.16E5	1.16E5	1.16E5	1.17E5	1.16E5±0.3%

Table 5.14 Metal core: comparison of reference toxicity characteristics

146

	Burner Reference	Burner Once-Thru	Burner Recycle	Partial Burner	Fissile Self sufficient
Conversion Ratio	0.45	0.46	0.45	0.67	1.04
Enrichment, TRU/HM	25.8	26.4	30.2	26.9	22.5
Burnup Swing,$\%\Delta k$	-5.9	-5.3	-4.3	-3.6	-3.5
TRU Consumption, kg/y	250	244	255	148	-10
Recycled TRU, kg/y	-	-	1502	1416	1316
Blending Ratio (Recycle/Total TRU Feed)	-	0.00	0.86	0.91	1.00
HM Loading, kg/y	-	5785	5833	8018	15,499
Delayed Neutron Fraction	-	3.51E-3	3.15E-3	3.25E-3	3.33E-3
Doppler Coefficient, $Tdk/dT \times 10^{-3}$	-	-1.46	-1.35	-1.51	-1.80
Sodium Void Worth * $\%\Delta k/kk'$					
Active Core	-	0.61	1.10	1.00	0.80
Above-Core Plenum	-	-0.80	-0.73	-0.73	-0.77
Total	-	-0.19	0.37	0.27	0.03

* for voiding of flowing sodium in the drivers only.

Table 5.15 Recycle benchmark neutronic performance characteristics comparison

	CR = 0.5 Once-Through	CR = 0.5 Recycle	CR = 0.75 Recycle	CR = 1.0 Recycle
Feed Stream, CD/y				
Uranium *	6.54E2	6.34E2	9.87E2	2.18E3
Transuranics	2.06E8	3.51E7	2.03E7	-
BOEC Inventory, CD				
Uranium	4.06E1	1.40E2	1.24E2	1.29E2
Transuranics	6.57E8	2.23E9	1.54E9	4.11E8
Short-lived FP	1.90E7	1.84E7	1.90E7	2.02E7
Long-lived FP	1.29E2	1.20E2	1.25E2	1.32E2
EOEC Inventory, CD				
Uranium	6.65E1	1.87E2	1.56E2	1.38E2
Transuranics	1.04E9	2.26E9	1.57E9	4.25E8
Short-lived FP	3.72E7	3.58E7	3.67E7	3.90E7
Long-lived FP	2.50E2	2.40E2	2.43E2	2.56E2
Recycled Feed, CD/y				
Transuranics	-	6.32E8	4.39E9	1.21E8
Waste Stream, CD/y				
Transuranics	3.22E8	6.32E5	4.39E5	1.21E5
Short-lived FP	1.53E7	1.48E7	1.53E7	1.62E7
Long-lived FP	1.03E2	1.00E2	1.02E2	1.07E2

* includes toxicity of equilibrium daughters (e.g., Pb-210), which are dumped in the mill tailings.

Table 5.16 Comparison of toxicity flows (all toxicities in Cancer Dose, CD)

Chapter 6

PLUTONIUM FUEL WITHOUT URANIUM

6.1 Introduction

Over the past 30 years many different fuel matrices have been investigated for plutonium fuels in various countries, including ceramic, metal, molten salt and others. Almost invariably, these different fuel matrices incorporated uranium as well as plutonium, in order to benefit from the production of Pu-239 from U-238 captures, which adds to the energy output for a given initial plutonium loading. Consequently, investigations have concentrated mainly on fuels such as $(Pu, U)O_2$, $(Pu, U)C$, $(Pu, U)N$, Pu/U/Zr and so on. More recently, ex-weapons plutonium has become available for destruction/ degradation which, combined with increasing stocks of civilian plutonium has shifted the emphasis to that of burning plutonium in reactors as a means of minimising the risks of proliferation. Consequently, new concepts and technologies are expected to be evaluated and proposed for the effective burning of plutonium in reactors.

From this perspective, the most effective way is adopting plutonium fuel without uranium. Although a considerable amount of research and development on fuels with uranium has been carried out, past work on non-uranic plutonium fuels is very limited. A plutonium fuel based on a non-uranic carrier must fit in with a practical fuel cycle technology, have good irradiation performance and be capable of being manufactured economically.

This paper summarises the present status of and the major issues associated with the research and development of plutonium fuels with a non-uranic carrier, concentrating in particular on the physics.

6.2 Research and development status of plutonium fuel without uranium

6.2.1 Summary of existing reactor studies

This section briefly summarises existing reactor studies and fuel studies of plutonium fuel without uranium.

Canada

Theoretical studies for the CANDU reactor type have been performed. So far, no tests of manufacturability or irradiation performance have been made.

The potential for disposition of weapons plutonium in the CANDU system has been studied, [1], ZrO_2 and BeO having been considered as inert matrices. The calculations show a plutonium annihilation rate of 2.5 kg /EFPD in a 680 MWe CANDU.

149

The lack of U-238 in the fuel and consequent increase in the Doppler coefficient is not expected to be a problem in the CANDU design. From the safety analysis perspective, in the current design the only accident that can quickly introduce an amount of reactivity close to prompt critical is a LOCA, and this is dealt with by two diverse shut down systems without much help from the negative fuel temperature coefficient. For fuels without U-238 the coolant void reactivity is in any event reduced to close to zero so that the problem is dealt with.

For the current natural uranium fuelled designs the fuel temperature coefficient is calculated to be about $-4 \cdot 10^{-6}$ k/°C. For the Pu without U fuel, the main effects with increasing fuel temperature are a decrease in η for Pu-239 due to an increase in the effective thermal neutron temperature and an increase in reactivity due to Doppler broadening of the Pu resonances. The net effect is calculated to be slightly negative: a fuel temperature coefficient of about $-2 \cdot 10^{-6}$ k/°C. This value has not been confirmed experimentally, but it is not expected that significant redesign of the control system will be required.

Additionally, the Th/U-233 fuel cycle has been studied in the CANDU system, the purpose of which is to avoid the production of plutonium and the higher actinides, [2].

France

An extensive programme of research and development into non-uranic fuels has been carried out and continues.

In the 1970s, a neptunium transmutation programme was carried out in the CELESTIN reactors with the aim of producing Pu-238 for pacemakers. Initially the neptunium was in the form of Np-Al alloy plate, and was later substituted by NpO_2/MgO. From 1973 to 1976, 26 target elements, each containing approximately 700 g of neptunium, were fabricated at Cadarache. These were then irradiated in the CELESTIN reactors for approximately 100 days in a flux of 2 to $3 \cdot 10^{13}$ n/cm^2/s. The conversion rate of neptunium into plutonium was around 6% with a Pu-238 isotopic abundance greater than 90%, [3] & [4].

In the heterogeneous concept of the ACTINEAU experiment, which is designed to demonstrate the feasibility of minor actinide (Am and Np) incineration in PWRs and to study the metallurgic behaviour of rods, the following matrices are selected based on cross-sections and material properties: $MgAl_2O_4$ and Al_2O_3 as reference materials, and CeO_2, Y_2O_3 and $Y_3Al_5O_{12}$ as alternatives, [3]. The experiment is scheduled to take place in the OPERA loop of the OSIRIS reactor in 1996, [4].

New theoretical and experimental work for fast reactors is in progress for cores with Pu fuel without U to enhance the Pu consumption up to the theoretical limit (~ 100 kg/TWhe). The studies carried out in this field in the frame of the CAPRA programme, [5], have been exploratory in character insofar as questions of principle had to be answered concerning the safety of these cores, a priori characterised by the absence of a Doppler effect before starting more detailed evaluations. The most notable result is that clearly there is no impossibility as regards safety (stability, flow transients, reactivity transients, core accidents) provided that a fuel design, meeting a certain number of constraints, can actually be put forward. At the present stage of the studies, some promising generic fuel families have been proposed but no clear and definitive choice has emerged since a specific experimental validation programme is needed.

One important lesson drawn from the studies was that the low Doppler effect does not pose any problem for flow transients insofar as it is compensated by the low sodium void effect resulting from the very high dilution of the uranium-free fuel in the core. However, to achieve acceptable behaviour in the reactivity transients besides requiring fuels with considerable margins as regards melting, it seems indispensable to find another prompt negative reactivity feedback. The studies showed that to a certain degree, axial thermal expansion of the fuel could ensure this function, or that certain neutron absorber materials such as tungsten for example could restore a Doppler effect but important uncertainties remain in the estimation of its absolute value. The studies revealed that these absorber materials have, as expected, a very negative impact on the sodium void reactivity and that only materials with a moderate neutron absorption cross-section could be tolerated. For these two situations, the requirement was to find a fuel solution in which any power rise would instantaneously put either the expansion or the Doppler reactivity effect into action. This supposes excellent thermal coupling between the Pu and the other constituent elements of the fuel. Moreover for the expansion effect, the design of the fuel must allow the latter to expand freely within its cladding.

In a process for which the implementation of a structural Doppler effect is desired, massive introduction of an absorber structural element would result in considerable increase of the positive core sodium void reactivity: to restore margins as regards this value, it seems advisable to consider smaller sized cores. As a consequence, two possible routes for further investigation (they can be mixed) seem to be either a large core in which an efficient axial expansion of the fuel is required or a smaller core in which a Doppler effect from structural materials is sought.

Several reference fuel candidates are:

- "cermets" like PuO_2 - Cr or PuO_2 - alloy based on (Cr, V, W) or W;
- "cercers" like PuO_2 - $MgAl_2O_4$ or MgO, or;
- solid solutions like (Pu, rare earths) O_2 or Pu (rare earths or Zr or Ti) N.

Experimental research and development work is under way and two irradiations have been launched in 1994/1995:

- *MATINA* – in the PHENIX reactor – [6] to study inert matrices such as MgO, $MgAl_2O_4$, TiN, Cr, V, Nb and W, and also cercers such as UO_2 - MgO and UO_2 - $MgAl_2O_4$.

- *TRABANT* – in the HFR reactor – where one pin is made of fuels without uranium: PuO_2 - MgO and (Pu, Ce)O_2.

Fabrication, dissolution and sodium compatibility tests have been carried out for most of the inert matrices mentioned above, [7].

Germany

A study has been carried out in which cerium was substituted for uranium in some PWR fuel assemblies, [8], with the intention of reducing the buildup of plutonium. In this study, fissile plutonium production was compared between (U, Pu)MOX fuels and (Ce, Pu)MOX fuels taking the portion of MOX fuel assemblies as a parameter. It was pointed out that the Doppler coefficient was negative but was too small in magnitude.

The utilisation of (Th, Pu)O_2 fuel in a standard PWR has been investigated in collaboration with Brazil and it has been concluded that nearly the same performance of the reactor could be expected using (Th, Pu)O_2 fuel as for UO_2 fuel in a standard PWR, [9]. It is noted that the development of (Th, Pu)O_2 fuel technology using cerium as a simulator material has made good progress; cerium has proved to be a good Pu-simulator since its chemical, ceramographic and physical properties are similar.

Japan

Survey calculations on a 800 MWe fast reactor core fuelled with PuO_2/Al_2O_3, PuO_2/BeO and $PuO_2/Al_2O_3/B_4C$ have been carried out in order to study not only the nuclear characteristics but also the safety characteristics.

A design study of a FBR safety research reactor, the driver of which core consists of PuO_2/BeO fuel, has been in progress, although this core was not originally intended for burning plutonium. In support of this work, a manufacturability test of UO_2/BeO pellets, in place of PuO_2/BeO is planned as a first step, [10].

An investigation for once through type fuel has been carried out aiming at burning weapons grade plutonium, [11]. $PuO_2/ThO_2/Al_2O_3$ and $PuO_2/ZrO_2(Gd, Y)/Al_2O_3$ are considered to be desirable from the point of view of the chemical properties of the materials, which would make it difficult to reprocess and extract plutonium.

In the burnup calculation study, it was estimated that as much as 83% of the total plutonium and 98% of Pu-239 is transmuted, and the quality of plutonium becomes very poor in the spent fuels. The void and Doppler reactivity coefficients are, however, very small in zirconia type fuel $(PuO_2/ZrO_2 (Gd,Y)/Al_2O_3)$ LWRs as compared with conventional LWR.

In fact, it was estimated that void reactivity is less than 1/5 of that of conventional PWR at beginning of equilibrium cycle (BOC), and becomes comparable at end of cycle (EOC). Both BOC and EOC, Doppler reactivity is smaller than those of PWR by the factor ranging from 4 to 7. The delayed neutron fraction is about 0.29% at BOC and increases to 0.35% at EOC.

The Doppler reactivity of the zirconia type fuel PWR can be improved with additives such as ThO_2, UO_2 or W. About 10 mol% of depleted UO_2 in the fuel can increase the Doppler reactivity about 4 times. Heterogeneous cores with PuO_2/UO_2 are also effective to increase this reactivity. In such cores, the delayed neutron fraction increases.

Switzerland

An assessment of the influence of different neutron absorbers (needed for reactivity compensation) on the core performance of PWRs has been started [12]. Experimental work is concentrated on the preparation of inert matrix materials like CeO_2, ZrO_2, Y, Al-Garnet, Spinell, TiN and ZrN co-precipitated with plutonium and the minor actinides neptunium and americium by a sol-gel process, [13].

United States

As part of the disposition study of weapons plutonium by fission, the advanced liquid metal reactor (ALMR) with Pu/Zr metal fuel, [14] & [15], and the modular high-temperature gas-cooled reactor with a plutonium core using PuO_2 particle fuel, [14] & [16], have been studied. The Pu/Zr fuel will require some development, though irradiation experience with reference Pu/U/Zr fuel is extensive.

The introduction of hafnium (Hf) as a diluent material (in the form of a mixture of binary Pu-Zr and Hf-Zr alloy), is proposed as one of the options to dispose weapons plutonium in IFR, [15], for which the following benefits are expected :

- The Hf acts as a burnable poison,
- The Hf resonance capture will provide negative Doppler reactivity.

This study concludes that non-uranic fuel results in unfavourable changes to the delayed neutron fraction, Doppler coefficient, and sodium void worth, in addition to large burnup reactivity loss.

$PuAl_4$ dispersed in aluminium and PuO_2 embedded in carbon and sealed in a silicon carbide shell (TRISO-coated PuO_2) have been examined and a Pu/Al composite with additional resonance absorbing poisons such as tungsten has been chosen for a newly developed low-temperature, low-pressure, low-power-density, and low-coolant-flow rate light water reactor concept that quickly burns weapons grade plutonium, [17].

Russia

Although only limited information is available, the utilisation of weapons grade plutonium in thermal and fast reactors is being investigated, [18]. Three types of reactor cores, a WWER core type, a conventional BN-800 core type, and a new BN-800 core based on a cermet fuel with metallic thorium and plutonium oxide, have been compared. It is concluded that the new BN-800 core has the shortest time for transformation of weapons grade plutonium into normal power plant plutonium and also yields the smallest radiotoxicity.

6.2.2 Core characteristics

This section illustrates the core performance of various non-uranic plutonium fuels, using the example of a 800 MWe fast reactor calculated by PNC :

Main specifications of the core

- Fuel .... PuO_2/Al_2O_3 (reference)
- Core height .... 60 cm
- Cycle length .... 228 EFPD
- Refuelling batch 6
- Volume fraction .... fuel/structure/sodium = 41/21/38
- Pu .... discharged from LWR.

	PuO$_2$/AlO$_3$ (reference)	PuO$_2$/BeO	PuO$_2$/Al$_2$O$_3$/B$_4$C (10%)
Burnup reactivity (% Δk/kk')	6.52	7.36	4.33
Power peaking factor (BOC/EOC)	1.91 / 1.60	1.94 / 1.72	1.54 / 1.51
Na void reactivity (EOC) (% Δk/kk')	-0.76	-0.37	+1.48
Doppler coefficient (10^{-3} Tdk/dT)	-3.4* (-1.3)**	-7.1*	-0.18*
β*eff* (x 10^{-3})	2.8	2.8	2.4
Pu fissile inventory (BOC/EOC) (10^3 kg)	4.6 / 4.1	4.0 / 3.5	7.6 / 7.1
Pu fissile consumption (kg/TWhe)	~ 110	~ 110	~ 110
Pu fissile consumption rate * (%)	50	56	33

* including both heavy metal and structural materials. Value in parentheses excludes structural material.
** (consumed Pu) / (loaded Pu).

*Table 6.1 **Core characteristics of plutonium fuel without uranium***

The distinctive features of the fast reactor core with non-uranic plutonium fuel are:

- Very large burnup reactivity compared with around 3% for the conventional PuO$_2$/UO$_2$ fuelled breeder reactor core;

- Large power peaking factor compared with around 1.5 of the conventional PuO$_2$/UO$_2$ fuelled fast breeder reactor core;

- Small and even negative sodium void reactivity compared with around 2% Δk/kk' of the conventional PuO$_2$/UO$_2$ fuelled fast breeder reactor core;

- Reduced Doppler coefficient compared with around -9.10^{-3} Tdk/dT of the conventional PuO$_2$/UO$_2$ fuelled fast breeder reactor core;

- Reduced delayed neutron fraction compared with around 3.8.10^{-3} of the conventional PuO$_2$/UO$_2$ fuelled fast breeder reactor core;

- PuO$_2$/BeO fuel improves the Doppler coefficient while it worsens sodium void reactivity and the burnup reactivity;

- The introduction of B$_4$C improves the burnup reactivity and the power peaking factor while it worsens the sodium void reactivity and the Doppler coefficient.

The major core characteristics described above will determine the kind of research and development that should be carried out in the field of reactor physics before the practical application of the fast reactor core with non-uranic plutonium fuel can be realised.

As for reactivity coefficients, a drastic reduction of the sodium void reactivity is necessary in order to compensate the low level of the Doppler effect. From this point of view, elements such as W, Mo and Nb lead to too large positive sodium void reactivity, while the use of quasi-transparent elements such as Al, Mg and Si makes it possible to obtain low enough sodium void reactivity. Between these two extremes, there exist some possible intermediate elements such as Cr and Ce, [19].

6.2.3 Fuel characteristics

The fuel characteristics are provided in Tables 6.2 to 6.4.

Desirable fuel characteristics for burning plutonium effectively in reactors

From the point of view of soundness under irradiation

- Solid solution of plutonium with other elements or plutonium compounds with other compounds, or stable compounds (eutectic structure, cermet types) for desired burnup;

- No phase changes of fuels over the temperature range to be used in a reactor. Phase changes are expected to result in problems, such as:
  - *volume change of fuels,*
  - *rapid gas release from fuels,*
  - *severe pellet - cladding mechanical interactions (PCMI),*

  which conventionally lead to fuel failures;

- Good compatibility of fuels with cladding and coolant materials;

- Good solubilities of fission products into fuels for higher burnup;

- Good mechanical and thermal properties of fuels:
  - *high melting point,*
  - *high thermal conductivity,*
  - *low swelling, proper plasticity considering PCMI,*
  - *low gas release,*
  - *others;*

From the fuel cycle technology point of view

- Fabrication

  - *applicability of those practical procedures which have been investigated experimentally so far,*
  - *less further requirement for R&D of fuels,*
  - *higher feasibility;*

- Reprocessing

  - *solubility in the case of recycling mode,*
  - *lack of solubility in the case of once-through mode;*

- Disposal

  - *good compatibility of spent fuel treated wastes with container materials,*
  - *low leaching rate of treated wastes from spent fuel, namely good resistance against leaching,*
  - *good stability of treated waste materials against radiation and in disposal conditions.*

Some surveys and evaluations of fuel without uranium

Oxides

Firstly, it is desirable that fuels with PuO_2 should be a solid solution. Partner oxides of PuO_2 should preferably be cubic and have a fluorite structure which is capable of solid solution formation. Partner oxides are expected to be ThO_2, CeO_2, HfO_2, ZrO_2 and TiO_2 from the point of view of the periodic law of elements; however, currently only ThO_2, [20], CeO_2, [20] and ZrO_2, [21] have been investigated in regard of the phase diagram and irradiation properties. ThO_2 and CeO_2 form good solid solutions, but ZrO_2 forms two phases (cubic + mono ZrO_2, above 78% ZrO_2). In addition, ZrO_2 is hard to dissolve in an acid solution. Therefore, PuO_2- ZrO_2 is considered to pose difficulties for reprocessing. Y_2O_3 is considered to form stable fuel with PuO_2. The partner oxides desirable for solid solution formation with PuO_2 are limited to a small number. Besides fuels forming the solid solution, MgO, Al_2O_3 and BeO, which form eutectic structure with PuO_2, are now under investigation. Eutectics have a tendency of making melting temperature lower than that of PuO_2. MgO [22] and Al_2O_3 and are soluble in acid while BeO is not.

Nitrides

PuN has a NaCl type crystal lattice and is capable of forming the solid solution with the same type of nitrides, [23], such as UN, ThN, CeN, NdN, HfN, ZrN, TiN and YN. Concerning PuN/UN, many investigations have been carried out as it is a promising fuel. However, other nitride fuels (with the exception of PuN/UN) have not been subject to research. In general, nitride fuels have many advantages in respect to good performance under irradiation, [24] and the fuel cycle technologies of fabrication and reprocessing. Nitride fuels are not suitable for LWR fuel as they react with H_2O.

Nitride fuels have higher melting points, higher thermal conductivities and produce smaller quantities of TRU than MOX. Since it is well known that the C-14 production has to be reduced to a minimum, enrichment in N-15 should be envisaged. Then, economic production of N-15 and the decomposition of the fuel at high temperature need to be considered. The key step is first of all to carry out irradiation examinations to evaluate the fuel performance.

Alloys

- Solid solution, [25]
  An important point of the solid solution is that the face-centered-cubic, delta-phase Pu should be stabilised to room temperature by alloying. Small additions of such elements as Al, Ce, Er, Dy, Hf, In, Sc, Zn and Zr are known to cause metastable retention of the delta solid solution under certain conditions, but only Al and Ce are known to stabilise this solid solution at room temperature. Pu dissolved in Zr forms a Pu solid solution which has been considered to be a possible reactor fuel;

- Intermetallic compounds (dispersion type fuels),
  The fuels considered are as follows:

  - *$PuAl_4$ in Al matrix, [26],reaching a Pu atom burnup of up to 60%,*
  - *Pu_2Cu_{11} in Cu matrix,*
  - *$PuFe_2$ in Fe matrix.*

Another important group of high melting intermetallic compounds includes oxides, carbides and nitrides. In particular, mixture and solid solution of PuO_2 with other materials have been studied most extensively.

6.3 Issues for future studies

Some calculational studies and experimental studies related to non-uranic plutonium fuels have been carried out already. However, these existing studies are very limited both in quality and in quantity. This is why it is essential that extensive surveys and evaluations should be continued.

The major issues for future studies are listed below :

Neutronics

- Doppler reactivity

  - *obtaining comprehensive basic data,*
  - *evaluation and reduction of uncertainties*
  - *design study of the practicability of a core with near zero Doppler coefficient;*

- Reduction of burnup reactivity (design study)

  - *survey the burnable poison → positive sodium void reactivity trade off,*
  - *survey to identify appropriate moderator;*

- Critical experiments

  - *validation of neutronics parameters,*
  - *evaluation of the impact of uncertainties relating to minor Pu isotopes,*
  - *nuclear data validation for inert matrices;*

- Core safety studies including transient analyses

  - *evaluation of the effects of reduced Doppler coefficient and β_{eff},*
  - *evaluation of the advantages of small and even negative sodium void reactivity;*

- Power peaking and power swing

  - *design study to mitigate power peaking and power swing;*

Fuel

- Theoretical and experimental studies on fuel

  - *phase studies on homogeneity of Pu (in solid solution) and heterogeneity of Pu phase (in eutectic structure, deposit and cermet) in the fuel matrix,*
  - *spot size of Pu phase and its homogeneous distribution in the fuel matrix,*
  - *assessment of the possible upper limit in plutonium enrichment,*
  - *data for fuel design and performance*
    physical, chemical and mechanical properties (thermal conductivity, thermal expansion, melting point, compatibility with cladding and coolant, etc.);

- Irradiation experiments to validate the in-pile behaviour

  - *phase study on homogeneity and spot size of Pu phase and fission product behaviour,*
  - *fuel performance on swelling, gas release, PCMI, etc.,*
  - *compatibility of fuel with cladding and coolant,*
  - *migration of Pu, minor actinides and fission products in fuel with a temperature gradient;*

- Fabrication

  - *mixing procedure to ensure good homogeneity of Pu in the fuel matrix, considering the spot size of Pu phase,*
  - *applicability tests of practical fabrication procedures;*

- Reprocessing

  - *dissolubility in acid for recycling mode,*
  - *feasibility studies with existing reprocessing processes.*

6.4 Conclusions

Only a limited number of investigations of non-uranic plutonium fuels have been conducted so far. In particular, little experience exists regarding manufacturability and irradiation.

Since the most suitable inert matrix differs depending on reactor type, reprocessing method, fuel cycle mode (once-through or recycle) and so on, it is necessary to select the proper inert matrix for the system, taking account of the nuclear characteristics, physical properties, irradiation performances and reprocessing abilities. It is recommended that manufacturability tests and irradiation tests for plutonium fuels with selected inert matrices should be given priority. Furthermore, reactor physics experiments and calculational studies to evaluate the nuclear characteristics of plutonium fuels with the inert matrices are also recommended to be pursued.

References

[1]   J. Pitre et al.: "The Role of CANDU reactor in the Disposition of Plutonium", ANS Topical meeting, Knoxville, 1984.

[2]   A. R. Dastur et al.: "The Role of CANDU in Reducing the Radiotoxicity of Spent Fuel", GLOBAL '93, Seattle, September 1993.

[3]   C. Prunier et al.: "First Results and Future Trends for the Transmutation of Long-lived Radioactive Wastes", Safewaste '93, Avignon, France, June 1993.

[4]   Y. Guerin et al.: "Transmutation of Minor Actinides in PWRs ; Preparation of ACTINEAU Experiment", Proc. GLOBAL '93, Seattle, September 1993.

[5]   J. Rouault: "Feasibility Assessment of the Plutonium (and Other Actinides) Fast Burners", Proc. 2nd Int. CAPRA Seminar, Karlsruhe, Sept. 21-22, 1994.

[6]   M. Salvatores, C. Prunier, Y. Guerin, A. Zaetta: "Transmutation Studies at CEA in the frame of the SPIN Program – Objectives, Results and Future Trends", Proc. Int. Conf. on Accelerator-driven Transmutation Technologies and Applications, Las Vegas, July 25-29 1994.

[7]   J. C. Chauvin (CEA): personal communication.

[8]   H. Schmieder et al.: "R&D Activities for Actinide Partitioning and Transmutation in the Nuclear Research Center Karlsruhe", Proc. International Information Exchange Meeting on "Actinide and Fission Product Separation and Transmutation", Mito, Japan, November 1990.

[9]   M. Peehs, G. Schlosser: "Prospect of Thorium Fuel Cycles in a Standard PWR", Siemens Forschungs- und Entwicklunsgs- Bericht, vol. 15 (1986) Nr.4.

[10]   K. Yamaguchi: "Design Study on the FBR Safety Research Reactor - Literature Survey on the Development of BeO Diluted Type Fuels", (in Japanese), PNC ZN 941092-006, January 1992.

[11]   H. Akie et al.: "A New Concept of Once-through Burning for Nuclear Warheads Plutonium", Proc. ICENESS '93, Makuhari, Japan, September 1993.

[12]   P. Wydler (PSI): personal communication.

[13]   G. Lederberger, F. Ingold, R. W. Stratton, H. P. Alder (PSI)
       C. Prunier, D. Warin, M. Bauer (CEA/DRN/DEC):
       "Application of Gel-Co-Conversion for TRU (Pu, Np, Am) Fuel and Target Preparation", Proc. International Conference GLOBAL '93, Seattle, September 1993.

[14] C. E. Walter, R. P. Omberg: "Disposition of Weapon Plutonium by Fission", GLOBAL '93, Seattle, September 1993.

[15] R. N. Hill et al.: "Physics Studies of Weapons Disposition in the IFR Closed Fuel Cycle", Proc. ANS Topical Meeting, Knoxville, 1994.

[16] D. Alberstein et al.: "The MHTGR - Maximum Plutonium Destruction without Recycle", transactions of ANS, San Francisco, pg. 94, November 1993.

[17] J. M. Ryskamp et al.: "New Reactor Concept without Uranium or Thorium for Burning Weapons-grade Plutonium", transactions of ANS, San Francisco, pg. 96, November 1993.

[18] V. M. Murogov et al.: "Pu/Th/U Cycle. Fuel Cycle Free from Plutonium", IAEA Meeting on Long-term Option for Pu Disposition, Vienna, January 1994.

[19] J. Rouault et al.: "Physics of Plutonium Burning in Fast Reactors: Impact on Burner Cores design", ANS Meeting, Knoxville, 1994.

[20] R. N. R. Mulford et al.: "PuO_2-CeO_2, PuO_2-ThO_2 solid solution", J. Phys. Chem., 62. 1466 (1958).

[21] D. F. Carroll: "PuO_2-ZrO_2 solid solution", J. Am. Ceramic Soc., 46. 194 (1963).

[22] D. F. Carroll: "PuO_2-MgO solid solution", J. Am. Ceramic Soc., 47 [12] 650 (1964).

[23] "Hand Book of Inorganic Chemistry, Uranium Suppl.", C7, Springerverlag, Berlin, 69 (1981).

[24] W. F. Lyon et al.: "Performance Analysis of a Mixed Nitride Fuel System for an Advanced Liquid Metal Reactor", Am. Nucl. Soc. Winter Meeting, Washington D.C., November 1990.

[25] O J. Wick et al.: "Plutonium Handbook, A Guide to the Technology", vol. 1, p. 193, Gordon & Breach, Science Publisher (1967).

[26] O J. Wick et al.: "Plutonium Handbook, A Guide to the Technology", vol. 1, p. 194, Gordon & Breach, Science Publisher (1967).

Table 6.2 Characteristics of plutonium without uranium fuels (oxide)

	Crystal structure	Phase change with temp.	Solubility of FP in fuel	Compatibility with cladding and coolant				Fuel cycle technology			Irradiation experiment
				SS	Zircaloy	Na	H$_2$O	fabrication	reprocessing	dissolubility in acid	
PuO$_2$-UO$_2$	good solid solution at least up to 45% PuO$_2$ [2]	2 phases occur 20~30% PuO$_2$	good but small metal and perovskite phase occur	good	good	reaction	reaction	many experiments, Pu-thermal, FBR	many experiments, Pu-thermal, FBR	good at least up to 35% PuO$_2$ [7]	many
PuO$_2$-ThO$_2$	possible solid solution [1]		probably similar above	probably good	probably good	probably reaction	probably reaction			possible depending on PuO$_2$ content	
PuO$_2$-CeO$_2$	possible solid solution [1]			probably good	probably good			some experiments	some experiments, experimental scale	possible depending PuO$_2$ content	
PuO$_2$-ZrO$_2$	good solid solution [2]	2 phases occur above 77% ZrO$_2$		good	good		probably stable	some experiments	probably impossible	ZrO$_2$ is insoluble PuO$_2$ is insoluble	few
PuO$_2$-Y$_2$O$_3$				probably good	probably good		probably good			Y$_2$O$_3$ is dissoluble	
PuO$_2$-BeO	eutectic [3], eutectic temp: 43% PuO$_2$-57%BeO (2135°C)	above 2135°C (BeO-PuO$_2$) liquid + BeO		good [3]	good [3]	probably good	probably stable	under study ~ 5% PuO$_2$-BeO	difficult	insoluble [3]	
PuO$_2$-MgO	eutectic [4], eutectic temp: 43% PuO$_2$-57%MgO (2000°C)	above 2000°C (MgO-PuO$_2$) liquid + MgO		good [6]		good	MgO is dissoluble	some experiments	difficult	PuO$_2$ is insoluble	under planning

	Crystal structure	Phase change with temp.	Solubility of FP in fuel	Compatibility with cladding and coolant				Fuel cycle technology			Irradiation experiment
				SS	Zircaloy	Na	H_2O	fabrication	reprocessing	dissolubility in acid	
PuO_2-$ZrO(Y_2O_3)$-Al_2O_3	PuO_2-$ZrO(Y_2O_3))S$ S = Al_2O_3 [5], eutectic			probably good	probably good			under study ~ 3% PuO_2	difficult	insoluble	under planning simulation study
PuN-UN	solid solution		good	good	good	good	reaction	use N-15	under study recovery of C-14	good	many
PuN-ThN	solid solution			probably good	probably good	probably reaction	reaction			good	
PuN-CeN	solid solution			probably good	probably good	probably reaction	reaction			good	
PuN-ZrN	solid solution			probably good	probably good	probably reaction	reaction			good	
PuN-Si_3N_4	dispersion type with Si_3N_4			probably good	probably good			use N-15			possible use below 1500°C
Pu-U-Zr				good		good		many	many pyro-chemical		many
Pu-Zr		many		good	good	good		few			few
Pu-W (or Mo)											

1   Mulford R.N.R. et al. "PuO_2-CeO_2, PuO_2-ThO_2 solid solution", J. Phys. Chem., 62.1466 (1958)

2   Carroll D.F. "PuO_2-ZrO_2 solid solution", J. Am. Ceramic Soc., 46.194(1963)

3   Yamaguchi K., "Design Study on the FBR Safety Research Reactor - Literature Survey on the Development of BeO Diluted Type Fuels" (in Japanese), PNC ZN 941092-006, Jan. 1992

4   Carroll D.F., "PuO_2- MgO System", J. Am. Ceramic Soc., 47 [13] 650 (1964)

5   Akie H. et al., "A New Concept of Once-through Burning for Nuclear Warheads Plutonium", Proc. ICENESS '93, Makuhari Japan, Sept. 1993

6   Paprocki S.J. et al., "The Chemical Reactions of PuO_2 with Reactor Materials", USAEC Report BMI-1580, May 1962

7   Rouault J. et al., "Physics of Plutonium Burning in Fast Reactors: Impact on Burner Cores Design", ANS Meeting, Knoxville, 1994

Table 6.3 *Properties of oxides*

	Crystal structure	Melting point	Thermal conductivity BTU (hr × ft ×°F)⁻¹ at 1000°C	Thermal expansion coefficient $\times 10^{-6}$ (in °F)⁻¹	Radiation damage
PuO₂	F.C.C. fluorite	2360°C [1]	1.22 [3]	10.9 (25 ~ 1000°C) [1]	
UO₂	F.C.C. fluorite	2860°C [1]	1.62 [1]	11.2 (25 ~ 1750°C) [1]	
ThO₂	F.C.C. fluorite	3200 +/- 100°C [1]	1.76 [1]	10.2 (25 ~ 1750°C) [1]	
CeO₂	cubic fluorite	2840°C [2]	0.7 [1]	7.9 (1400 ~ 1920°F) [1]	
HfO₂	cubic (stabilised with CaO, MgO or Y₂O₃)	2538°C [2]		5.24 (80 ~ 1800°C) [2]	
ZrO₂	cubic (stabilised with CaO, MgO or Y₂O₃)	2766°C [2]	1.33 [2]	8.50 (1400 ~ 2400°F) [2]	stable
TiO₂	tetragonal	1838°C [2]	2.3 [2]	4.90 (80 ~ 2730°F) [2]	
Y₂O₃	cubic	2410°C [2]	1.8 [2]	5.2 (1200 ~ 2780°F) [2]	
BaO	cubic	1923°C [2]			
SrO	cubic	2455°C [2]		8.0 (1200 ~ 2200°F) [2]	
CaO	cubic	2614°C [2]	4.5 [2]	8.0 (1000 ~ 2000°F) [2]	
MgO	cubic	2852°C [2]	3.9 [2]	8.75 (1000 ~ 1800°F) [2]	stable
BeO	α hexagonal β tetragonal	2570°C [2]	12 [2]	5.71 (800 ~ 2200°F) [2]	unstable above 10^{20} n/cm² [1] He production in matrix
αAl₂O₃	hexagonal above 1000°C	2049°C [2]	3.5 [2]	6.6 (1800 ~ 3200°F)	
γAl₂O₃	cubic below 1000°C				

[1] "Introduction to Ceramic Material and Technologies" (in Japanese), edited by Industry Technology Center, 1979

[2] Lynch J.E. et al., "Engineering Properties of Ceramics - Data Book to Guide Material Selections", BMI, Ohio, 1969

[3] Conti A. (CEA): personal communication

163

Table 6.4 Properties of nitrides

	Crystal structure	Melting point	Thermal conductivity BTU $(hr \times ft \times F)^{-1}$ at 1000°C	Thermal expansion coefficient $x10^{-6}$ (in F°)$^{-1}$	Radiation damage
PuN	cubic [2]	2749°C [2]	7.5 [4]	6.83 (70 ~ 1650°F) [2]	
UN	cubic [2]	2904°C [2]	14.0 [4]	5.48 (68 ~ 2730°F) [2]	
ThN	cubic [2]	2788°C [2]		4.10 (65 ~ 1430°F) [2]	
CeN	cubic [1]	2980°C [3]			
HfN	cubic [1]	3300 ~ 3307°C [1]	9.0 [2]	3.6 (75 ~ 2500°C) [2]	
ZrN	cubic [1]	2930 ~ 2980°C [1]	12.4 [2]	4.3 (1100 ~ 2600°F) [2]	
TiN	cubic [1]	2900 ~ 3220°C [1]	16.0 [2]	5.0 (1100 ~ 2600°F) [2]	
NbN	cubic [1]	2050°C [1]			
AlN	hexagonal [2]	2200 ~ 2300°C [1]	12.0 [2]	3.16 (212 ~ 1830°F) [2]	
Si3N4	hexagonal [1]	decomposition above 1900°C [1]	3.8 [2]	1.4 (70 ~ 1800°F) [2]	
YN	cubic	2671°C [2]			

[1] "Introduction to Ceramic Material and Technologies" (in Japanese), edited by Industry Technology Center, 1979

[2] Lynch J.E. et al., "Engineering Properties of Ceramics - Data Book to Guide Material Selections", BMI, Ohio, 1969

[3] Massalski T.B., "Binary Alloy Phase Diagrams", 2nd edition ASM International, 1990

[4] Arai et al., "Dependence of the Thermal Conductivity of (U,Pu)N on Porosity and Plutonium Content", Journal of Nuclear Materials 195, (1992)

Chapter 7

RECYCLING OF PLUTONIUM IN ADVANCED CONVERTER REACTORS

7.1 Introduction

In the past the general belief in the nuclear industries of the world was that recycle of Pu would be necessary in the latter years of the 20th century to make sufficient use of the uranium resource to satisfy the needs of mankind. It was anticipated that this would mean that widespread deployment of fast breeder reactors would take place. To provide Pu for the start-up of these reactors plans were made, and in some countries implemented, to reprocess thermal reactor spent fuel. In some countries it was considered that there might be justification for the development of thermal reactors which could more efficiently burn recycled Pu, thus extending the world's uranium resource. Such reactors could not achieve the breeding offered by fast reactors, but also did not require the development of radically new technologies for their implementation. Such reactors may be categorised as Advanced Converter Reactors, examples would be the heavy water moderated reactors: the Japanese ATR, variants of the CANDU, and variants of the Siemens pressure vessel type PHWR, in addition one could include: variants of the HTGR, and advanced PWR designs.

7.2 The Japanese ATR

The ATR is a heavy water moderated, boiling light water cooled pressure tube type reactor designed specifically to burn plutonium. A prototype version (Fugen 165 MWe) has been in operation since 1979 and has demonstrated successful operation with both enriched uranium and MOX fuel operation to burnups of 20 MWd/kg. About 1000 kg of fissile Pu from the initial loading has been consumed in Fugen. The fuel cycle was closed for this reactor when in 1988 MOX fuel assemblies containing Pu from reprocessed Fugen fuel were placed in the reactor. Plans now call for a 600 MWe demonstration plant to be built for operation early in the next century. Fuel burnup for this demonstration plant is planned to be higher at about 38 MWd/kg.

The ATR offers the following advantages over recycling Pu in an LWR: somewhat more energy is derived from the recycled Pu (for a fuel with fissile content of 3.1% an average burnup of 30 MWd/kg is predicted), and the sensitivity of the burnup to the Pu isotopic composition is much less than in an LWR because of the more thermalised neutron spectrum.

The development programme associated with the Fugen reactor has produced a considerable amount of data against which nuclear data and calculational methods have been validated. The data include those derived from many zero energy measurements in the DCA facility as well as those obtained from operation of Fugen itself. In DCA values of k-effective as well as many reaction rate ratios were measured for fuels having three different Pu contents. Measured values were compared with values calculated by the WIMS-ATR code; a version of the WIMS-D code with a specially modified library. In general the calculational results agreed with the measurements within about 15% for the reaction rate ratios, while k-effective agreed within about 0.1% k [1,2,3]. From Fugen, plutonium isotopic ratios have been measured in spent fuel and compared with WIMS-ATR calculations with an ENDF/B-IV data set [4,5]. Agreement was good. Also from Fugen values of k-effective for various start up cores were measured as were values for the first eight cycles. Again good agreement was obtained [6].

7.3 The CANDU reactor

The CANDU reactor achieves a very high neutron economy by using heavy water as moderator and as coolant in a pressure tube design that allows on power refuelling with a rather simple fuel bundle. These properties combine to provide the potential for considerable fuel cycle flexibility [7]. There have been many studies indicating the feasibility of burning MOX fuel in CANDU, with the Pu derived either from CANDU or LWR spent fuel [8,9,10]. In the past it was thought that deriving Pu from CANDU spent fuel would be economic, despite the rather low Pu content in the fuel (about 0.37%). This has not proven to be the case and emphasis has shifted to cycles that are synergistic with the LWR, the so-called Tandem cycles. The Tandem cycle denotes the recycle in CANDU of both the uranium and plutonium recovered from spent LWR fuel. However, in a broader sense the Tandem cycle can refer to the recycle of the plutonium alone, with the uranium component of the MOX fuel coming from either: the reprocessed fuel (having a U-235 content of about 0.9%), depleted uranium from enrichment plant tails (0.2% U-235), or even natural uranium. The recycled material could be recovered using conventional reprocessing technology or through a simpler cheaper and more proliferation-proof process in which the Pu is not separated from the uranium, so-called co-processing.

The main advantage of recycling the LWR Pu and U through a CANDU is the increased energy obtained compared to recycle through an LWR. As much as twice the energy can, in principle, be obtained [10] while the fissile content of the fuel is reduced to levels so low that further reprocessing is not economic.

Manufacture of CANDU MOX fuel has been demonstrated on a laboratory scale in the Recycle Fuel Fabrication Laboratory (RFFL) at Chalk River Laboratories (CRL) [11], as has its irradiation in the NRU research reactor.

A series of zero energy physics tests have been performed in the ZED-2 critical facility at CRL with 37-element CANDU type MOX fuel bundles. These have been compared with WIMS-AECL calculations and the agreement found to be satisfactory [12,13].

System studies of the burning of MOX fuel in CANDU have also been performed [9]. These include studies of: refuelling schemes, reactivity of control and safety absorbers, fuel handling, safety analyses, and many other important factors. The studies indicate the feasibility of adapting the CANDU reactor for plutonium fuel cycles.

In 1992 AECL, KAERI and the US Department of State, completed Phase I of an assessment of using PWR spent fuel directly in CANDU without reprocessing or use of any wet chemical process. This class of synergistic fuel cycles is referred to as "DUPIC", for *D*irect *U*se of spent *P*WR Fuel *I*n *C*ANDU [14]. The use of only dry processes enhances the safeguardability of this fuel cycle. Two broad classes of direct use options were considered: mechanical reconfiguration and powder processing.

The study concluded that one of the powder processing options, OREOX (for Oxidation, Reduction of Enriched Oxide fuel) was most promising, largely because of the homogeneity of the resultant powder and pellets. One advantage of this option is that it removes a high fraction of volatile and gaseous fission products, thereby improving fuel burnup. The CANDU burnup with the OREOX option is predicted to be about 18 MWd/kg, using spent fuel from the reference Korean PWR, which has an average discharge burnup of 35 MWd/kg (initial U-235 enrichment of 3.25%). Despite removal of some of the fission products the gamma activity of the fuel remains high and it is clear that remote manufacturing techniques in a shielded facility will be required. The safeguards assessment concluded that the proliferation risks of the DUPIC cycle are minimal, and that presently known safeguards systems and technologies can be modified or adapted to meet safeguards requirements.

The programme is now continuing with optimisation of the OREOX process, fabrication of fuel elements and bundles from spent PWR fuel for subsequent irradiation testing and post irradiation examination, development of remote fabrication techniques and development of appropriate safeguards technology.

From the point of view of the physics of the DUPIC cycle there do not appear to be any impediments to its implementation, it is clear, however that much development of other technologies is required.

7.4 Pressure vessel type heavy water reactor (PHWR)

The Siemens pressure vessel type heavy water reactor is capable of recycling plutonium by spiking fuel assemblies. This allows an economical utilisation of plutonium through the use of a loading scheme with high local plutonium content and without penalising local power peaking, [15]; full core length fuel assemblies can in fact be shuffled from the core edge to inner positions. This form of spiking was tested at the Karlsruhe MZFR reactor in 1972 by inserting eight MOX fuel assemblies.

7.5 High conversions LWRs

The main purpose of utilising high conversion reactors to burn Pu is to increase uranium utilisation. The conversion ratio of LWRs can be increased by increasing the fuel to moderator ratio of the lattice. In this way it is in principle possible to increase uranium utilisation by a factor of about four if the fuel-to-moderator ratio is increased by about the same amount [16]. To achieve much higher values of the fuel to moderator ratio than those in the current PWRs will require a change to a hexagonal cell. There have been many studies of high conversion light water reactors, the results of many of them being summarised in Reference [17]. With regard to core physics calculation validation data, Reference 16 reports results of programs in France and Switzerland [18,19].

The data base provided by the Swiss/German experimental programme on LWHCR test lattices [20,21] is broad, in terms of both the range of high converter design characteristics represented and the types of integral data measured. Changes in moderation ratio and effective fuel enrichment were covered – with investigation of neutron balance components, moderator voidage effects, influence of lattice poisoning, relative control rod effectiveness and power depression/peaking due to core heterogeneities. In general, comparisons of measured and calculated lattice parameters reported to date have shown that discrepancies in k-infinity predictions can be reduced to the experimental error limit by applying improved calculational methods and cross-section sets. This is not yet the case for some of the measurable reaction rate ratios, indicating that compensation of errors may be occurring. Remaining deficiencies in nuclear data and/or resonance treatment thus require further attention.

In France, within the framework of a co-operation between the French electricity supplier, EDF, the reactor designer Framatome, and the French Atomic Commission, CEA, an exhaustive experimental programme was undertaken in the centre of the EOLE facility at Cadarache to measure the fundamental neutronic parameters involved in HCLWR design calculations.

Two types of lattice were investigated: a very tight one, called ERASME/S (Mod. Ratio = 0.5) and a more realistic one, ERASME/R (Mod. Ratio = 0.9) in order to be representative of the Framatome HCLWR projects. Calculations are in good agreement with experimental values [24]. Void configurations have also been investigated in both lattices, where the reactivity effect was negative because of the very large effect of axial leakage. Calculations are in good agreement with the experimental results.

In parallel, in two similar undermoderated lattices placed in the zero power MINERVE facility and in the 8 MWth MELUSINE core, neutronic parameters related to the change in reactivity with burnup (the reactivity effect of fission products, and capture cross-sections of the main heavy isotopes) have been measured [25]. Calculational results show some discrepancies for capture cross-sections but the fission product reactivity effect seems to be well estimated [26].

7.6 Conclusions

Studies and demonstrations have shown that heavy water moderated reactors can burn the Pu and U produced from reprocessing LWR spent fuel. Such reactors generally require a lower mass of fissile material to maintain criticality and therefore can burn their fuel to lower fissile content than an LWR. Thus more energy can be extracted from the reprocessed materials and their fissile content reduced to the point where final disposal may be an economically attractive option.

There appear to be no limitations imposed by lack of knowledge of the neutronics and physics of such reactors, but there may be challenges in other areas. As an example, such challenges are being addressed in the DUPIC programme where Canada, Korea and the USA are studying a dry reprocessing route for recycling LWR spent fuel through a CANDU reactor.

References

[1] T. Wakabayashi et al.: "Thermal Neutron Behaviour in Cluster-Type Plutonium Fuel Lattices" Nucl. Sci & Eng., 63 (1977) 292.

[2] N. Fukumura: "Measurement of Local Power Peaking Factors in Heavy-Water Moderated Plutonium Lattices", J. Nucl. Sci. Technol. 18 (1981) 285.

[3] S. Sawai et al.: "Characteristics of Plutonium Utilization in the Heavy-Water Moderated Boiling Light Water Cooled Reactor ATR", Nucl. Eng. & Design, 125 (1991) 251.

[4] R. Yamanaka et al.: "Post Irradiation Examination of Fugen MOX Fuel", PNC Report, PNCT N341 84-48 (1984).

[5] N. Kawata, K. Shimomura: "MOX Fuel Utilization in ATR", PNC Report PNCT N3410 87-010 (1987).

[6] Y. Shiratori et al.: "Core Management of the MOX Fuel Loaded Heavy Water Reactor", International Conference on the Physics of Reactors: Operation, Design and Computation (Physor '90), Marseille, 1990.

[7] D.F. Torgerson, P.G. Boczar, A.R. Dastur: "CANDU Fuel Cycle Flexibility", 9th Pacific Basin Nuclear Conference, Sidney, Australia, 1994

[8] J.B. Slater: "CANDU Advanced Fuel Cycles - A Long Term Energy Source", AECL Report, AECL-1901, 1986.

[9] J.R. Hardman, A.R. Dastur: "CANDU Nuclear Power Plant Optimization with MOX Fuel", International Conference on Design and Safety of Advanced Nuclear Power Plants, Tokyo, 1992.

[10] P.G. Boczar, I.J. Hastings, A. Celli: "Recycling in CANDU of Uranium and/or Plutonium from Spent LWR Fuel", AECL Report, AECL-10018, 1989

[11] T.J. Carter: "The Recycle Fuel Fabrication Laboratory at Chalk River", Proc. First Intl. Conf. CANDU Fuel, Canadian Nuclear Society, 1986.

[12] R.T. Jones, J. Griffiths, A. Okazaki: "Reactor Physics Measurements in Support of Advanced Fuel Cyclers in CANDU Reactors", AECL Report, AECL-10004, 1989.

[13] R.T. Jones, J. Griffiths: "Studies in the ZED-2 Critical Facility of Reactivity Coefficients for CANDU Cores Fuelled with Plutonium Fuels", International Conference on the Physics of Reactors: Operation, Design and Computation (Physor '90), Marseille, 1990.

[14] H. Keil, P.G. Boczar, H.S. Park: "Options for the Direct Use of Spent PWR Fuel in CANDU (DUPIC)", Ed. P.G. Boczar, Third International Conference on CANDU Fuel, Chalk River, 1992.

[15] F. Dusch: "Advanced Fuel Utilization in Heavy Water Reactors with Slightly Enriched Fuel and Plutonium Spiking", Kerntechnik, 50 (1987) 249.

[16] C.A. Götzmann, H. Markl, H. Moldaschl: "Rational for Pressurized Water High Conversion Reactor (PWHCR) Development Strategy", in Technical Aspects of High Converter Reactors, IAEA-TECDOC-638, 1992.

[17] "Technical Aspects of High Converter Reactors", Proceedings of a Technical Committee Meeting, Nuremberg, 1990, IAEA-TECDOC-638, 1992.

[18] J.L. Nigon, J. Mondot: "Experimental Support to Tight Lattice and Plutonium Core Studies", in Technical Aspects of High Converter Reactors, IAEA-TECDOC-638, 1992.

[19] R. Chawla, H. D. Berger, H. Hager, R. Seiler: "The PROTEUS Phase II Experiments of Data Base for LWHCR Physics Validation", in Technical Aspects of High Converter Reactors, IAEA-TECDOC-638, 1992.

[20] Chawla R. et al.: "Analysis of PROTEUS Phase II Experiments in Support of Light Water High Conversion Reactor Design", Kerntechnik 57 (1992) No. 1.

[21] R. Böhme et al.: "Neutron Balance Investigations for a High Enrichment MOX-LWR Lattice under Normal and Voided Conditions", Proc. of International Reactor Physics Conference, Tel Aviv, Israel, 1994.

[22] A. Santamarina, L. Martin-Deidier, S. Cathalau et al.: "ERASME : an Extensive Experiment for LWCHR Design Qualification", Top. Meet. on Reactor Physics, Saratoga Springs, Sept 1986.

[23] A. Santamarina, L. Martin-Deidier, S. Cathalau et al.: "Undermoderated PWR Neutronic Qualification through the ERASME Experiments", Top. Meet. on Advances in Reactor Physics and Mathematical Computation, Paris, April 27-30, 1987.

[24] S. Cathalau: "Validation of Basic Nuclear Data for HCLWR Using Zero Power Critical Experiments", Proc. Top. Meet. on Advance in Reactor Physics, Vol. 2 pp 385/396, Charleston SC, USA, March 1992.

[25] M. Darrouzet, J. Bergeron et al.: "The French Program Addressing the Requirements of Future Pressurized Water Reactors", Nuc. Tech. Vol. 80/2, pp. 269/281 1988.

[26] P. Chaucheprat, J. Mondot et al.: "MORGANE/S : Fission Product Capture Measurements in a HCLWR Tight Lattice", Top. Meet. on Reactor Physics, Jackson Hole USA, Sept 1988.

Chapter 8

THE USE OF RECYCLED URANIUM

8.1 Introduction

For typical thermal reactor burnups only about 4% of the initial heavy metal is converted from uranium to fission products or other actinides. Thus the decision to reprocess spent fuel from thermal reactors necessarily results in the production of quantities of uranium that are large compared with the masses of plutonium and fission products also produced. Consequently the possible disposition options for this recovered uranium (RU) deserve to be discussed at the same time as those of the plutonium, especially when one considers that this uranium often has a significant fissile content.

It is not anticipated that there would be any technical limitations which prevent or make uncertain the utilisation of the RU in any of the ways described below. The decisions of the owners of RU will rather be based on economic arguments, somewhat colored by their opinions of the likelihood of there being problems with any of the proposed routes for disposition.

8.2 The resource

Most of the fuel that has been and will be reprocessed comes from LWRs (both PWRs and BWRs, but mainly PWRs). The fissile content of the resulting uranium depends both on the reactor type and the burnup of the fuel. For standard PWRs burning the fuel to about 33 MWd/kg the RU contains about 0.9% U-235 [1], the figure is smaller for RU from higher burnup PWRs and for BWRs, in fact in [2] it is stated that the residual value of uranium separated from BWR spent fuel is lower than that of natural uranium.

In the United Kingdom there is experience with reprocessing Magnox reactor spent fuel, which contains only about 0.45% U-235, and with using large quantities of the resulting uranium. Up to 1987 about 15000 tU had been enriched to produce 1500 tU of recycle AGR fuel.

In all RU there has been some build up of U-236 and some reduction in U-234 due to neutron capture during the irradiation, this changes the neutronic properties of the uranium for further use and must be properly accounted for in whatever disposition is considered. There is also a buildup of U-232 in the irradiation, although only to very low levels. Nevertheless, because decay of the U-232 leads to some emitters of very hard gamma rays, it must be considered from a radiological point of view; if only to show that at the levels pertaining there need be no concern.

8.3 Means of utilising recovered uranium

There are, broadly speaking, at least four possibilities for utilising recovered uranium:

- Re-enrichment to the level required for re-irradiation in an LWR or other reactor requiring enriched fuel (such as an AGR);

- Re-fabrication for irradiation in a reactor which can economically burn the RU without it requiring re-enrichment, such CANDU or the ATR;

- Mixing with Pu to manufacture MOX fuel for burning in a suitable reactor; and

- Storage for possible re-enrichment by a highly selective laser process at some later date when such a possibility is available on an industrial scale.

8.4 Re-enrichment

Re-enrichment by both the gaseous diffusion and the centrifuge processes has been successfully demonstrated, although the centrifuge is generally regarded as the more suitable because the lower holdup of material in the process allows the batches of uranium to remain better separated thus reducing the problem of contamination with U-234 and U-236 and other isotopes such as Tc-99 [3]. In re-enrichment the lighter isotopes are concentrated more rapidly. Thus the U-232 and U-234 become larger fractions of the enriched material relative to the U-235 than they were in the source, whereas the U-236 is relatively reduced. This can lead to certain difficulties: the further concentration of the U-232 increased its radiological significance, and both U-234 and U-236 being neutron absorbers, it is necessary to "over-enrich" relative to material produced from natural uranium to produce fuel that will achieve the same burnup.

Some of the earliest experience was obtained in the United Kingdom where uranium recovered from reprocessed Magnox fuel has for many years been enriched from its initial U-235 content of about 0.45% to that required for AGR fuel. In this case the concentrations of the minor isotopes are small enough that the enriched material can be treated radiologically as though sourced from natural uranium. The quantities of U-234 and U-236 are low enough that the "over-enrichment" penalty is also low, especially in the soft neutron spectrum of the AGR where the importance of U-236 absorption is less significant.

In the case of re-enrichment for reuse in LWRs, experience has also been satisfactory. Several countries have already demonstrated good results or have ongoing programmes [3, 4, 5, 6]. For use in LWRs, where the required enrichment is higher and the spectrum harder, the required "over enrichment" is also higher, but has proved easy to allow for by means of a simple formula. It is also necessary to observe some extra precautions during processing due to the slightly increased radiological hazard, but these have not proven onerous. For example Urenco in its centrifuge enrichment plant in the Netherlands is routinely re-enriching recovered uranium for various clients for use in their LWRs. Only part of the plant is used for this purpose and only limited areas give concern from a radiological hazard point of view. Change of the plant to this mode of operation was achieved with minimal interruption of normal operation.

8.5 Burning without re-enrichment

Another option is to burn the recovered uranium without re-enrichment. This can be done economically in a reactor designed to burn low enrichment fuel. This possibility has been studied for the CANDU reactor [7] where it was calculated that RU from an LWR containing 0.9% U-235 could be burned to at least 13 MWd/kg. The soft neutron spectrum of the CANDU is advantageous in achieving these burnups since it minimises the absorbing effect of the U-236. A joint feasibility study with Cogema, in which LWR derived RU was pressed and sintered into CANDU pellets indicated no radiological problems with fuel manufacture [8]. Burning RU in this way would allow about twice the energy to be extracted from it compared to re-enrichment and burning in an LWR, while at the same time the U-235 content is reduced to that typical of enrichment tails, thus preparing the fuel for final disposal.

8.6 Burning as MOX

This option, in which the RU is mixed with Pu to achieve the required fissile content, is technically feasible. However, if the target reactor is an LWR this option cannot deal with all the RU because the Pu fraction required in the MOX fuel is higher than that in the spent fuel. This option is not popular with LWR owners because the balance of the RU still has to be dealt with and more MOX fuel has to be manufactured to utilise all the Pu. High converter reactors such as CANDU or the Japanese ATR could have a role to play in burning MOX fuel consisting of a mixture of all the Pu and RU.

8.7 Storing

The final possibility is to store the RU for a possible future in which its use is made more appealing by changing economic circumstances or by technical developments such as highly selective inexpensive laser enrichment.

8.8 Conclusions

There are a number of choices available for the utilisation of the growing quantity of recovered uranium in the world. The efficacy of the various options has been demonstrated to various degrees and there do not appear to be any scientific of technical issues that would preclude use of any of the methods. The owners of recovered uranium will decide on the basis of their own situations which route is most appropriate for them.

References

[1]   "Plutonium Fuel: An Assessment", Report by an Expert Group, NEA OECD, 1989.

[2]   Nuclear Fuel, 1994 Feb 14, Page 14.

[3]   J.L. Guillet, J.C. Guyot, G. Manet: "Uranium Recycling In Water Reactors", in Recycling of Plutonium and Uranium in Water Reactor Fuels, IAEA Publication IWGFPT/35, 1990.

[4]   K. Takai, S. Abeta, S. Hattori, A. Maru: "Reprocessed Uranium Fuel Fabrication in Japan", in Recycling of Plutonium and Uranium in Water Reactor Fuels, IAEA Publication IWGFPT/35, 1990.

[5]   W.D. Krebs, G.J. Schlosser: "Status of Design and Operational Experience with Enriched Reprocessed Uranium (ERU) in the Federal Republic of Germany", in Recycling of Plutonium and Uranium in Water Reactor Fuels, IAEA Publication IWGFPT/35, 1990.

[6]   M. Galimberti, M. Pontiq: "The EDF Strategy of Uranium and Plutonium Recycling in PWRs", in Recycling of Plutonium and Uranium in Water Reactor Fuels, IAEA Publication IWGFPT/35, 1990.

[7]   P.G. Boczar, I.J. Hastings, A. Celli: "Recycling in CANDU of Uranium and/or Plutonium from Spent LWR Fuel", in Recycling of Plutonium and Uranium in Water Reactor Fuels, IAEA Publication IWGFPT/35, 1990.

[8]   P.G. Boczar et al.: "Recovered Uranium in CANDU: A Strategic Opportunity", Proceedings of the International Nuclear Congress and Exhibition (INC93), Toronto, 1993.

Chapter 9

CONCLUSIONS AND RECOMMENDATIONS

The conclusions and recommendations emerging from the analysis and discussion in the previous chapters are summarised by relevant issue.

9.1 Physics and some engineering aspects of plutonium recycling in light water reactors

Multirecycle of plutonium in LWRs offers a practical near term option for extracting further energy from LWR spent fuel and reprocessed plutonium.

Multirecycling plutonium in PWRs of current design can have limitations and related physics issues have been considered, in particular the plutonium content limitation to avoid positive void effects and the minimisation of minor actinide production during multirecycling. The present conclusions indicate a good understanding of physics, even if to establish a consensus on this item would need further scenario type of studies including lattice optimisation (e.g., under-moderated lattices). Experimental validation of integral parameters (reactivity coefficients) in the case of degraded plutonium isotopic composition may also be needed.

Some of the principal conclusions – essential for multiple recycling – are as follows.

1. Regarding nuclear design calculations, modern methods incorporating rigorous resonance self-shielding and modern nuclear data libraries such as JEF-2, ENDF/B VI or JENDL-3 are essential. Some data improvement, e.g., higher plutonium isotope data in the resonance region, etc., might be needed;

2. Appropriate and well tested calculation methods are widely available and should be used;

3. The interaction of neighbouring UO_2 and MOX elements should be investigated in more detail to obtain a clearer understanding of the results observed;

4. The optimal control of high burnup LWR cores (which may include burnable poisons) should be examined further;

5. Additionally, questions of material damage in the case of high burnup need to be clarified as well as other engineering constraints;

6. Sophisticated (three-dimensional) core calculations are recommended to study the void effect and could also be the object of a future benchmark exercise, possibly in relation with the analysis of an experiment to be made available to the international community;

7. At present, it appears that plutonium recycling in high burnup LWR cores can be performed twice without modifying current LWR designs;

8. Future experimental verification related to maximum plutonium content in the case of degraded plutonium isotopic composition is needed in clean lattice configurations with different moderator-to-fuel ratios and for possible void experiments with different leakage components. A co-ordinated international effort in this field would be highly beneficial;

9. For unchanged lattices the limit on plutonium content is in the range $12 \pm 1\%$;

10. Changes in the lattice in the sense of higher moderation can be foreseen to minimise the buildup of higher actinides and to increase the limits on plutonium content;

11. Multiple recycling of plutonium with high burnup (e.g., 50 MWd/kg) can have limitations due to considerations such as the buildup of Pu-238 and Pu-242 or the existence of positive reactivity feedback effects on complete coolant voiding at high plutonium contents or the increase of the buildup of higher actinides (Am, Cm, etc.). A specific future benchmark in this field could help in obtaining an international consensus on these limitations.

9.2 Plutonium recycling in fast reactors

Even if for the long term, the best use of plutonium is still in fast breeder reactors, fast burner reactor/LWR symbiosis offers potential for nuclear waste reduction by further extraction of energy from the multicycled LWR spent fuel and reprocessed plutonium.

Burner fast reactor physics benchmarks display a larger spread in results among participants than has been experienced for more conventional breeder designs.

1. Further efforts are warranted to assure that fast burner reactor designs of high transuranic content, plutonium of poor isotopic quality, and high leakage fraction can be accurately calculated;

2. Critical experiments should be considered to validate integral design parameters (such as reactivity coefficients);

3. Nuclear data of higher actinide isotopes which are relevant to waste heating rates and long-term waste radiotoxicity hazard should be refined (i.e., Cm isotopes data). An international agreement should be reached on the quality (data uncertainties and their impact) of these nuclear data. A recommendation in this direction should be made to the Nuclear Science Committee Working Party on Evaluation Co-operation.

9.3 Toxicity inventory and flow

As far as the toxicity inventory and flow are concerned, the parametric analysis with variation of the fast reactor conversion ratio reveals several features:

1. For a Once-Through fuel cycle, the toxicity in the waste stream is dominated by the transuranic component at discharge and this component maintains its toxicity for many thousands of years. In contrast by efficiently recycling of the transuranics back to the reactor, the short-lived fission products can be made to dominate the waste stream toxicity and the short-lived toxicity of fission products will decay away in several hundred years leading to significant reduction in stream toxicity consigned to the geological repository for the long-term waste;

2. Thus, by a symbiosis of LWR discharge supplying feedstock to fast burner designs, the toxicity outflow from the fast reactor cycle to the waste steam will be significantly lower than the inflow of toxicity in the feed stream from the LWR (which otherwise would be destined for waste). For the conversion ratio CR = 0.5 the toxicity is reduced by roughly half the introduction rate each year, and for the CR = 0.75 design, the toxicity is reduced by roughly a quarter each year at time of discharge. Loss of transuranics to the waste stream was assumed to be 0.1% / recycle;

3. Within 500 years the waste toxicity reduction has grown to a two orders of magnitude reduction as a result of removal and recycle of the transuranics;

4. If the loss of transuranics to the waste stream in the recycle chemistry were reduced by a further factor of 15 from the 0.1% / recycle value used in the benchmarks, then after 500 years, the radiotoxicity in the waste is less than that of the uranium ore which was mined to produce the fuel in the first place. The toxicity inventory in the reactor is nearly two orders of magnitude higher than the yearly flows of feed and waste toxicity, and is dominated by the transuranic component;

5. The in-core toxicity is significantly lower (factor of 5 times) in the CR = 1.0 and Once-Through system than in the burner designs due to multiple recycle isotopic changes which depend on the core conversion ratio; low conversion ratios lead to a buildup of higher actinides (with their associated higher toxicity levels) upon multiple recycle whereas fissile break even designs lead to a Pu-239 rich transuranic composition upon multiple recycle;

6. The results obtained in the oxide fuelled benchmark and in the first case of the metal-fuelled benchmark show similar trends as far as radiotoxicity inventory at the end of a single cycle is concerned. This fact indicates that, with the same hypothesis of high efficiency in the transuranic recovery at reprocessing (e.g., 0.1% loss to the waste stream), most of the conclusions related to the metal-fuelled multiple recycle benchmark (including the effects of the parametric conversion ratio variation) would be applicable to oxide-fuelled cores.

9.4 Plutonium fuel without uranium

Only a limited investigation on the plutonium fuels without uranium has been conducted so far. Especially, the experience of manufacturability and irradiation is limited.

Since the preferred inert matrix differs depending on reactor type, reprocessing method, fuel cycle mode (once-through or recycle) and so on, it is necessary to select the proper inert matrix for the system taking account of nuclear characteristics, physical properties, irradiation performances and reprocessing abilities. It is recommended to accelerate the manufacturability tests and the irradiation tests for plutonium fuels with the selected inert matrices. Furthermore, reactor physics experiments and calculational studies to evaluate the nuclear characteristics of the plutonium fuels with inert matrices are also recommended to be pursued. The problem of prompt reactivity feedback (e.g., Doppler) availability has to be considered in detail. Preliminary safety studies have indicated several possibilities to overcome potential difficulties in this field.

9.5 Recycle of plutonium through advanced converter reactors

Studies and demonstrations have shown that heavy water moderated reactors can burn the plutonium and U produced from reprocessing LWR spent fuel. Such reactors generally require a lower mass of fissile material to maintain criticality and therefore can burn their fuel to lower fissile content than LWRs. Thus more energy can be extracted from the reprocessed materials and their fissile content reduced to the point where final disposal may be an economically attractive option.

There appear to be no limitations imposed by lack of knowledge of the neutronics and physics of such reactors, but there may be challenges in other areas. As an example, such challenges are being addressed in the DUPIC program where Canada, Korea and the U.S.A. are studying a dry reprocessing route for recycling LWR spent fuel through a CANDU reactor.

9.6 Recycling of uranium

There are a number of choices available for the utilisation of the growing quantity of recovered uranium in the world. The efficacy of the various options has been demonstrated to various degrees and there do not appear to be any scientific or technical issues that would preclude use of any of the methods. The owners of recovered uranium will decide on the basis of their own situations which route is most appropriate for them.

9.7 Criticality safety and MOX fuel

Finally, it should be mentioned that criticality safety issues related to irradiated MOX fuel with high burnup should be the object of specific assessments in the frame of the NEA Nuclear Science Committee Group on Criticality Safety. These issues have not been addressed explicitly in the present report, but can have relevance in the future fuel cycle studies involving MOX fuels.

Acknowledgements

We would like to acknowledge here the permission for reproducing Figure 4.7 on page 108 [*H. Sekimoto, "Physics of Future Equilibrium State of Nuclear Energy Utilization" – Proceedings of Conference on Reactor Physics and Reactor Computations, Tel Aviv, 23-26 January 1994, Y. Ronen and E. Elias, eds, Ben Gurion University of the Negrev Press (1994)*] and Figure 4.8 on page 109 [*L. Koch, "Formation and Recycling of Minor Actinides in Nuclear Power Stations" – Handbook on the Physics and Chemistry of Actinides, Volume 4, Chapter 9, A. Freeman and C. Keller, eds, Elsevier Science Publishers B.V. (1986)*] and thank their respective publisher for it.

Appendix 1

Members of and Participants to
the NSC Working Party on the Physics of Plutonium Recycling

BELGIUM

D'HONDT, Pierre
MINSART, Georges

Centre d'Étude de l'Énergie Nucléaire (**SCK/CEN**)
200 Boeretang
2400 MOL

DERAMAIX, Paul
MALDAGUE, Thierry

Belgonucléaire
4 av. Ariane
1200 BRUXELLES

CANADA

JONES, Richard T.

AECL Research
Chalk River Nuclear Laboratories
CHALK RIVER
Ontario K0J 1J0

FRANCE

CATHALAU, Stephane
DASILVA, M.
FINCK, Phillip J.
GARNIER, Jean-Claude
MAGHNOUJ, A.
MARIMBEAU, Pierre
RAHLFS, Stefan
RIMPAULT, Gérald
SALVATORES, Massimo
VARAINE, F.

CEA
C.E. **Cadarache**
13108 ST. PAUL LEZ DURANCE Cedex

COSTE, Mireille
PUILL, André
SOLDEVILA, Michel
TELLIER, Henri

CEA
C.E. **Saclay**
91191 GIF SUR YVETTE Cedex

179

BARBRAULT, P.
VERGNES, Jean

Électricité de France (**EDF**)
Direction Etudes et Recherches
1,Avenue du Général de Gaulle
BP.408
92141 CLAMART Cedex

LEFEBVRE, Jean-Claude

Électricité de France (**EDF**)
SEPTEN
12-14 Avenue Dutrievoz
69628 VILLEURBANNE Cedex

AIGLE, Richard
KOLMAYER, André
POINOT, Christine

Société **Framatome**
EPN
Tour FIAT - Cedex 16
1, Place de la Coupole
92084 PARIS LA DEFENSE

GERMANY

BERNNAT, Wolfgang
KÄFER, S.
LUTZ, Dietrich
MATTES, Margarete

Universität Stuttgart
Institut für Kernenergetik und Energiesysteme (**IKE**)
Postfach 801140
70550 STUTTGART 80

HETZELT, Lothar
SCHLOSSER, Gerhard

SIEMENS AG/KWU BT14
Postfach 3220
Bunsenstr. 43
91050 ERLANGEN

KÜSTERS, Heinz
WIESE, Hans-Werner

Institute for Neutron & Reactor Physics
Kernforschungszentrum Karlsruhe (**KFK**)
Postfach 3640
76021 KARLSRUHE

ITALY

LANDEYRO, Pedro

ENEA - DRI/COMB
CRE Casaccia
S.P. Anguillarese, 301
00100 ROMA A.D.

JAPAN

ISHII, Kazuya
MARUYAMA, Hiromi

Energy Research Laboratory
Hitachi Ltd.
1168 Moriyama-cho,
Hitachi-shi, Ibaraki-ken 316

AKIE, Hiroshi
KANEKO, K.
MORI, T
OKUMURA, Keisuke
TAKANO, Hideki

Dept. of Reactor Engineering
JAERI
TOKAI-MURA, Naka-gun
Ibaraki-ken 319-11

IKEGAMI, Tetsuo
KOIZUMI, M.
OHKI, S.
WAKABAYASHI, Toshio
YAMAMOTO, Toshihisa

Power Reactor & Nuclear Fuel Development Corp. (**PNC**)
4002,Narita-cho, O-arai-machi
Higashi-Ibaraki-gun, 311-13

SAJI, Etsuro

TODEN Software, Inc.
In-core Fuel Management Syst.
6-19-15 Sinbashi
Minato-ku
TOKYO, 105

KAWASHIMA, Masatoshi
UENOHARA, Yuji
YAMAOKA, M.

Nuclear Engineering Laboratory
TOSHIBA Corporation
4-1, Ukishima-cho
Kawasaki-ku,
Kawasaki 210

KOREA (REPUBLIC OF)

JOO, Hyung-Kook

Korea Atomic Energy Research Institute (**KAERI**)
Advanced LWR Fuel Development
P.O. Box 105 Yusong-gu
Daeduk-Danji - TAEJON 305-600

NETHERLANDS

AALDIJK, J.K.
FREUDENREICH, W.E.
HOGENBIRK, Alfred
LI, J.M.
WICHERS, Victor A.

B.U. Nuclear Energy
ECN
Postbus 1
1755 ZG PETTEN

RUSSIAN FEDERATION

TSIBULIA, Anatoly

Institute of Physics and Power Engineering (**IPPE**)
249 020 OBNINSK
Kaluga Region

SWEDEN

EKBERG, Kim

Studsvik AB
Fack
611 82 NYKOEPING

SWITZERLAND

CHAWLA, Rakesh
HOLZGREWE, Frank
PARATTE, Jean-Marie
PELLONI, Sandro
WYDLER, Peter

Paul Scherrer Institute (PSI)
5232 VILLIGEN PSI

UNITED KINGDOM

HESKETH, Kevin
MANGHAM, G.

British Nuclear Fuels plc (BNFL)
Springfields
Preston
Lancashire PR4 0XJ

SMITH, P.

AEA Technology Winfrith
DORCHESTER
Dorset DT2 8DH

UNITED STATES OF AMERICA

BLOMQUIST, Roger N.
GRIMM, K.
HILL, Robert N.
PALMIOTTI, G.
WADE, David C.

Argonne National Laboratory (ANL)
9700 South Cass Avenue
ARGONNE, IL 60439

INTERNATIONAL ORGANIZATIONS

RINEJSKI, Anatoli

Division of Nuclear Power
International Atomic Energy Agency (IAEA)
P.O. Box 100
A-1400 WIEN

SARTORI, Enrico

OECD/Nuclear Energy Agency (NEA)
Le Seine-Saint Germain
12 boulevard des Iles
F-92130 ISSY-LES-MOULINEAUX

Symbols and abbreviations

ALI	Annual Limit on Intake
AGR	Advanced Gas-cooled Reactor
ALMR	Advanced LMR
APM	Atelier Pilote Marcoule
APOLLO	French reactor-cell code system
APWR	Advanced PWR
at%	per cent in atoms
ATR	Japanese Advanced Thermal Reactor
BOC (BOEC)	Beginning Of Cycle
BOL	Beginning Of Life
Bq	becquerel, standard radiation activity unit
BR	Breeding Ratio
BWR	Boiling-Water Reactor
°C	degrees Celsius
CANDU	*CAN*adian *D*euterium *U*ranium
CAPRA	*C*onsommation *A*ccrue de *P*lutonium dans les *RA*pides
CD/g **CD/y**	Cancer dose per gram Cancer dose per year
CEC	Commission of the European Communities
Ci	curie, radiation activity unit $1 \text{ Ci} = 3.7 \times 10^{10}$ Bq (becquerel)
COCA	MOX fabrication process at French CEA
CR	Conversion Ratio
EFPD	equivalent full power days

ENDF	evaluated nuclear data file (U.S.A.)
EOC (EOEC)	End Of Cycle
EOL	End Of Life
η	number of neutrons released per neutron absorbed
eV	electronvolt
FA	Fuel Assembly
FBR	Fast Breeder Reactor
FP	Fission Product
FR	Fast Reactor
GRANULAT	MOX fabrication process of Russia
GWe **GWth** **GWy**	gigawatt electric gigawatt days of thermal energy gigawatt year
HCLWR	High Conversion LWR = LWHCR
HFR	High-Flux Reactor
HLW	High-Level Waste
HLLW	High-Level Liquid Waste
HTGR	High-Temperature-Gas Reactor
HWR	Heavy-Water Reactor
IBG	Internal Breeding Gain
IFR	Integral Fast Reactor
ILW	Intermediate-Level Waste
ILLW	Intermediate-Level Liquid Waste
JEF	Joint Evaluated Files
JENDL	Japan Evaluated Nuclear Data Library
k	neutron multiplication factor
K	degrees Kelvin
kg/TWhe	kilogram per terawatt hours electric
kg/y	kilogram per year
KONVOI PWRs	series of 3 identical German PWRs
KRITZ	Critical facility at Studsvik, Sweden
LLFP / LFP	Long-lived Fission Products

LLW	Low-Level Waste
LMR	Liquid-Metal Reactor
LOCA	Loss Of Coolant Accident
LWHCR	Light-Water High-Conversion Reactor = HCLWR
LWR	Light-Water Reactor
MA	Minor Actinide
MAGNOX	*MAGN*esium *OX*ide
MELOX	French *MEL*ange *OX*ide (MOX) Plant
MeV	megaelectronvolt
MIMAS	MOX fabrication process at Belgonucléaire, Belgium

MOX
  H-MOX
  L-MOX
  M-MOX
	*M*ixed *OX*ide (uranium and plutonium)
	  MOX with high Pu content
	  MOX with low Pu content
	  MOX with medium Pu content

MWd/kg	megawattdays per kilogram, refers to heavy metal (HM) = GWd/t
MWe	megawatt electric
ϕ n/cm^{2}s	neutron flux
NEACRP	Nuclear Energy Agency Committee on Reactor Physics
n/f	neutrons per fission
NJOY	Data processing system
NSC	Nuclear Science Committee
OCOM	Optimised co-milling MOX fabrication process of Siemens, Germany
OECD/NEA	OECD Nuclear Energy Agency
OMR	Open-Market Recycling
OREOX	Oxidation, Reduction of Enriched Oxide fuel
PCMI	Pellet-Cladding Mechanical Interactions
pcm	10^{-5}
PHENIX	French fast reactor
PHWR	Pressurised Heavy-Water Reactor

PIE	Post Irradiation Examination
ppm	part per million (10^{-6})
PROTEUS	Swiss research reactor at PSI
Pu	plutonium
Pu$_{fiss}$	fissile plutonium
Pu$_{tot}$	total plutonium
PUREX	*Plutonium URanium EXtraction*
PWR	Pressurized-Water Reactor
PYRØ	Pyrometallurgy-based reprocessing
RSIC	Radiation Shielding Information Center
RU	Recovered Uranium
σ	standard deviation
SGR	Self-Generated Recycle
SLFP / SFP	Short-lived Fission Products
SMP	Sellafield MOX Plant
S$_n$	discrete ordinates radiation transport method
T	temperature
TRU	transuranium elements
TWhe	terawatt hours electric
U	uranium
UO$_2$	uranium dioxide = UOX
VIP	Venus International Programmes
WIMS	U.K. reactor-cell code system
w/o	weight % = wt%
WPPR	Working Party on the Physics of Plutonium Recycling
WWER	Russian Water-Water Power Reactor
y	year

Uranium and plutonium buildup and decay chains [1]

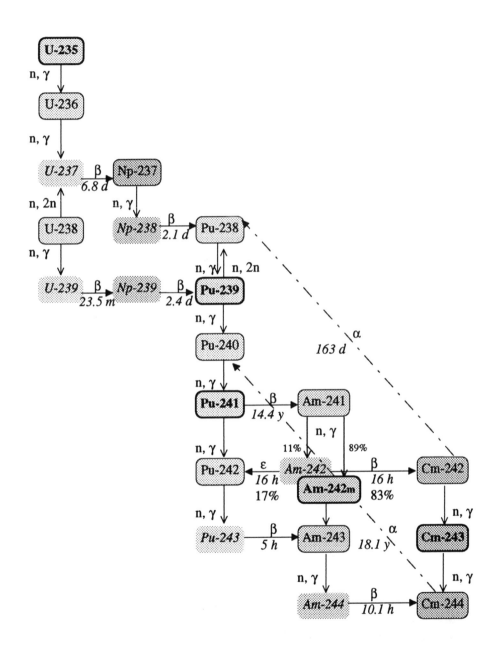

---

[1] for half-lives less than 20 years

Physics constants of radionuclides relevant for this Study

NUCLIDE	SPIN PARITY	HALF LIFE	Q-VALUE (EV)+	DECAY MODE	NATURAL ABUNDANCE
92-U-232	0+	69.80 y	5413.62	α,SF*	
92-U-233	5/2+	1.59E+05 y	4908.90	α,SF*	
92-U-234	0+	2.46E+05 y	4857.90	α,SF*	0.0055%
92-U-235	7/2-	7.04E+08 y	4679.00	α,SF*	0.7200%
92-U-236	0+	2.34E+07 y	4573.00	α,SF*	
92-U-237	1/2+	6.75 d	519.40	β	
92-U-238	0+	4.47E+09 y	4270.30	α,SF*	99.2745%
92-U-239	5/2+	23.47 m	1262.90	β	
93-Np-237	5/2+	2.14E+06 y	4957.30	α	
93-Np-238	2+	2.12 d	1291.90	β	
93-Np-239	5/2+	2.36 d	721.50	β	
94-Pu-238	0+	87.70 y	5593.27	α,SF*	
94-Pu-239	1/2+	2.41E+04 y	5243.50	α,SF*	
94-Pu-240	0+	6.56E+03 y	5255.90	α,SF*	
94-Pu-241	5/2+	14.40 y	20.81	β,α*	
94-Pu-242	0+	3.74E+05 y	4983.10	α,SF*	
94-Pu-243	7/2+	4.96 h	582.00	β	
95-Am-241	5/2-	432.70 y	5637.94	α,SF*	
95-Am-242	1-	16.02 h	661.20	β (0.827)	
			747.70	ε (0.173)	
95-Am-242m	5-	141.00 y	48.63	IT (0.995)	
			5366.70	α (0.005),SF*	
95-Am-243	5/2-	7.36E+03 y	5438.70	α,SF*	
95-Am-244	6-	10.10 h	1429.00	β	
96-Cm-242	0+	162.94 d	6215.76	α,SF*	
96-Cm-243	5/2+	30.00 y	8.00	ε (0.002)	
			6167.40	α (0.998)	
96-Cm-244	0+	18.10 y	5901.80	α,SF*	
96-Cm-245	7/2+	8.50E+03 y	5623.30	α	

+ *total decay energy available in the decay process (omitted for very small branching ratios).*
In the case of isomeric states it represents the energy of that state.
For β or ε decays it represents the energy corresponding to the mass difference between the initial and final atoms.

* *branching ratio smaller than 0.0001, SF = spontaneous fission*

Reference: JEF-2.2 Radioactive Decay Data, OECD/NEA, JEF Report 13, August 1994

MAIN SALES OUTLETS OF OECD PUBLICATIONS
PRINCIPAUX POINTS DE VENTE DES PUBLICATIONS DE L'OCDE

ARGENTINA – ARGENTINE
Carlos Hirsch S.R.L.
Galería Güemes, Florida 165, 4° Piso
1333 Buenos Aires   Tel. (1) 331.1787 y 331.2391
Telefax: (1) 331.1787

AUSTRALIA – AUSTRALIE
D.A. Information Services
648 Whitehorse Road, P.O.B 163
Mitcham, Victoria 3132   Tel. (03) 873.4411
Telefax: (03) 873.5679

AUSTRIA – AUTRICHE
Gerold & Co.
Graben 31
Wien I   Tel. (0222) 533.50.14
Telefax: (0222) 512.47.31.29

BELGIUM – BELGIQUE
Jean De Lannoy
Avenue du Roi 202
B-1060 Bruxelles   Tel. (02) 538.51.69/538.08.41
Telefax: (02) 538.08.41

CANADA
Renouf Publishing Company Ltd.
1294 Algoma Road
Ottawa, ON K1B 3W8   Tel. (613) 741.4333
Telefax: (613) 741.5439
Stores:
61 Sparks Street
Ottawa, ON K1P 5R1   Tel. (613) 238.8985
211 Yonge Street
Toronto, ON M5B 1M4   Tel. (416) 363.3171
Telefax: (416)363.59.63
Les Éditions La Liberté Inc.
3020 Chemin Sainte-Foy
Sainte-Foy, PQ G1X 3V6   Tel. (418) 658.3763
Telefax: (418) 658.3763

Federal Publications Inc.
165 University Avenue, Suite 701
Toronto, ON M5H 3B8   Tel. (416) 860.1611
Telefax: (416) 860.1608

Les Publications Fédérales
1185 Université
Montréal, QC H3B 3A7   Tel. (514) 954.1633
Telefax: (514) 954.1635

CHINA – CHINE
China National Publications Import
Export Corporation (CNPIEC)
16 Gongti E. Road, Chaoyang District
P.O. Box 88 or 50
Beijing 100704 PR   Tel. (01) 506.6688
Telefax: (01) 506.3101

CHINESE TAIPEI – TAIPEI CHINOIS
Good Faith Worldwide Int'l. Co. Ltd.
9th Floor, No. 118, Sec. 2
Chung Hsiao E. Road
Taipei   Tel. (02) 391.7396/391.7397
Telefax: (02) 394.9176

CZECH REPUBLIC – RÉPUBLIQUE TCHÈQUE
Artia Pegas Press Ltd.
Narodni Trida 25
POB 825
111 21 Praha 1   Tel. 26.65.68
Telefax: 26.20.81

DENMARK – DANEMARK
Munksgaard Book and Subscription Service
35, Nørre Søgade, P.O. Box 2148
DK-1016 København K   Tel. (33) 12.85.70
Telefax: (33) 12.93.87

EGYPT – ÉGYPTE
Middle East Observer
41 Sherif Street
Cairo   Tel. 392.6919
Telefax: 360-6804

FINLAND – FINLANDE
Akateeminen Kirjakauppa
Keskuskatu 1, P.O. Box 128
00100 Helsinki
Subscription Services/Agence d'abonnements :
P.O. Box 23
00371 Helsinki   Tel. (358 0) 12141
Telefax: (358 0) 121.4450

FRANCE
OECD/OCDE
Mail Orders/Commandes par correspondance:
2, rue André-Pascal
75775 Paris Cedex 16   Tel. (33-1) 45.24.82.00
Telefax: (33-1) 49.10.42.76
Telex: 640048 OCDE

Orders via Minitel, France only/
Commandes par Minitel, France exclusivement :
36 15 OCDE

OECD Bookshop/Librairie de l'OCDE :
33, rue Octave-Feuillet
75016 Paris   Tel. (33-1) 45.24.81.81
(33-1) 45.24.81.67

Documentation Française
29, quai Voltaire
75007 Paris   Tel. 40.15.70.00
Gibert Jeune (Droit-Économie)
6, place Saint-Michel
75006 Paris   Tel. 43.25.91.19
Librairie du Commerce International
10, avenue d'Iéna
75016 Paris   Tel. 40.73.34.60
Librairie Dunod
Université Paris-Dauphine
Place du Maréchal de Lattre de Tassigny
75016 Paris   Tel. (1) 44.05.40.13
Librairie Lavoisier
11, rue Lavoisier
75008 Paris   Tel. 42.65.39.95
Librairie L.G.D.J. - Montchrestien
20, rue Soufflot
75005 Paris   Tel. 46.33.89.85
Librairie des Sciences Politiques
30, rue Saint-Guillaume
75007 Paris   Tel. 45.48.36.02
P.U.F.
49, boulevard Saint-Michel
75005 Paris   Tel. 43.25.83.40
Librairie de l'Université
12a, rue Nazareth
13100 Aix-en-Provence   Tel. (16) 42.26.18.08
Documentation Française
165, rue Garibaldi
69003 Lyon   Tel. (16) 78.63.32.23
Librairie Decitre
29, place Bellecour
69002 Lyon   Tel. (16) 72.40.54.54
Librairie Sauramps
Le Triangle
34967 Montpellier Cedex 2   Tel. (16) 67.58.85.15
Tekefax: (16) 67.58.27.36

GERMANY – ALLEMAGNE
OECD Publications and Information Centre
August-Bebel-Allee 6
D-53175 Bonn   Tel. (0228) 959.120
Telefax: (0228) 959.12.17

GREECE – GRÈCE
Librairie Kauffmann
Mavrokordatou 9
106 78 Athens   Tel. (01) 32.55.321
Telefax: (01) 32.30.320

HONG-KONG
Swindon Book Co. Ltd.
Astoria Bldg. 3F
34 Ashley Road, Tsimshatsui
Kowloon, Hong Kong   Tel. 2376.2062
Telefax: 2376.0685

HUNGARY – HONGRIE
Euro Info Service
Margitsziget, Európa Ház
1138 Budapest   Tel. (1) 111.62.16
Telefax: (1) 111.60.61

ICELAND – ISLANDE
Mál Mog Menning
Laugavegi 18, Pósthólf 392
121 Reykjavik   Tel. (1) 552.4240
Telefax: (1) 562.3523

INDIA – INDE
Oxford Book and Stationery Co.
Scindia House
New Delhi 110001   Tel. (11) 331.5896/5308
Telefax: (11) 332.5993

17 Park Street
Calcutta 700016   Tel. 240832

INDONESIA – INDONÉSIE
Pdii-Lipi
P.O. Box 4298
Jakarta 12042   Tel. (21) 573.34.67
Telefax: (21) 573.34.67

IRELAND – IRLANDE
Government Supplies Agency
Publications Section
4/5 Harcourt Road
Dublin 2   Tel. 661.31.11
Telefax: 475.27.60

ISRAEL
Praedicta
5 Shatner Street
P.O. Box 34030
Jerusalem 91430   Tel. (2) 52.84.90/1/2
Telefax: (2) 52.84.93
R.O.Y. International
P.O. Box 13056
Tel Aviv 61130   Tel. (3) 49.61.08
Telefax: (3) 544.60.39
Palestinian Authority/Middle East:
INDEX Information Services
P.O.B. 19502
Jerusalem   Tel. (2) 27.12.19
Telefax: (2) 27.16.34

ITALY – ITALIE
Libreria Commissionaria Sansoni
Via Duca di Calabria 1/1
50125 Firenze   Tel. (055) 64.54.15
Telefax: (055) 64.12.57
Via Bartolini 29
20155 Milano   Tel. (02) 36.50.83
Editrice e Libreria Herder
Piazza Montecitorio 120
00186 Roma   Tel. 679.46.28
Telefax: 678.47.51
Libreria Hoepli
Via Hoepli 5
20121 Milano   Tel. (02) 86.54.46
Telefax: (02) 805.28.86
Libreria Scientifica
Dott. Lucio de Biasio 'Aeiou'
Via Coronelli, 6
20146 Milano   Tel. (02) 48.95.45.52
Telefax: (02) 48.95.45.48

JAPAN – JAPON
OECD Publications and Information Centre
Landic Akasaka Building
2-3-4 Akasaka, Minato-ku
Tokyo 107   Tel. (81.3) 3586.2016
Telefax: (81.3) 3584.7929

KOREA – CORÉE
Kyobo Book Centre Co. Ltd.
P.O. Box 1658, Kwang Hwa Moon
Seoul   Tel. 730.78.91
Telefax: 735.00.30

MALAYSIA – MALAISIE
University of Malaya Bookshop
University of Malaya
P.O. Box 1127, Jalan Pantai Baru
59700 Kuala Lumpur
Malaysia   Tel. 756.5000/756.5425
                            Telefax: 756.3246

MEXICO – MEXIQUE
Revistas y Periodicos Internacionales S.A. de C.V.
Florencia 57 - 1004
Mexico, D.F. 06600   Tel. 207.81.00
                            Telefax: 208.39.79

NETHERLANDS – PAYS-BAS
SDU Uitgeverij Plantijnstraat
Externe Fondsen
Postbus 20014
2500 EA's-Gravenhage   Tel. (070) 37.89.880
Voor bestellingen: Telefax: (070) 34.75.778

**NEW ZEALAND
NOUVELLE-ZÉLANDE**
Legislation Services
P.O. Box 12418
Thorndon, Wellington   Tel. (04) 496.5652
                            Telefax: (04) 496.5698

NORWAY – NORVÈGE
Narvesen Info Center – NIC
Bertrand Narvesens vei 2
P.O. Box 6125 Etterstad
0602 Oslo 6   Tel. (022) 57.33.00
                            Telefax: (022) 68.19.01

PAKISTAN
Mirza Book Agency
65 Shahrah Quaid-E-Azam
Lahore 54000   Tel. (42) 353.601
                            Telefax: (42) 231.730

PHILIPPINE – PHILIPPINES
International Book Center
5th Floor, Filipinas Life Bldg.
Ayala Avenue
Metro Manila   Tel. 81.96.76
                            Telex 23312 RHP PH

PORTUGAL
Livraria Portugal
Rua do Carmo 70-74
Apart. 2681
1200 Lisboa   Tel. (01) 347.49.82/5
                            Telefax: (01) 347.02.64

SINGAPORE – SINGAPOUR
Gower Asia Pacific Pte Ltd.
Golden Wheel Building
41, Kallang Pudding Road, No. 04-03
Singapore 1334   Tel. 741.5166
                            Telefax: 742.9356

SPAIN – ESPAGNE
Mundi-Prensa Libros S.A.
Castelló 37, Apartado 1223
Madrid 28001   Tel. (91) 431.33.99
                            Telefax: (91) 575.39.98

Libreria Internacional AEDOS
Consejo de Ciento 391
08009 – Barcelona   Tel. (93) 488.30.09
                            Telefax: (93) 487.76.59

Llibreria de la Generalitat
Palau Moja
Rambla dels Estudis, 118
08002 – Barcelona
                (Subscripcions) Tel. (93) 318.80.12
                (Publicacions) Tel. (93) 302.67.23
                            Telefax: (93) 412.18.54

SRI LANKA
Centre for Policy Research
c/o Colombo Agencies Ltd.
No. 300-304, Galle Road
Colombo 3   Tel. (1) 574240, 573551-2
                            Telefax: (1) 575394, 510711

SWEDEN – SUÈDE
Fritzes Customer Service
S–106 47 Stockholm   Tel. (08) 690.90.90
                            Telefax: (08) 20.50.21

Subscription Agency/Agence d'abonnements :
Wennergren-Williams Info AB
P.O. Box 1305
171 25 Solna   Tel. (08) 705.97.50
                            Telefax: (08) 27.00.71

SWITZERLAND – SUISSE
Maditec S.A. (Books and Periodicals - Livres
et périodiques)
Chemin des Palettes 4
Case postale 266
1020 Renens VD 1   Tel. (021) 635.08.65
                            Telefax: (021) 635.07.80

Librairie Payot S.A.
4, place Pépinet
CP 3212
1002 Lausanne   Tel. (021) 341.33.47
                            Telefax: (021) 341.33.45

Librairie Unilivres
6, rue de Candolle
1205 Genève   Tel. (022) 320.26.23
                            Telefax: (022) 329.73.18

Subscription Agency/Agence d'abonnements :
Dynapresse Marketing S.A.
38 avenue Vibert
1227 Carouge   Tel. (022) 308.07.89
                            Telefax: (022) 308.07.99

See also – Voir aussi :
OECD Publications and Information Centre
August-Bebel-Allee 6
D-53175 Bonn (Germany)   Tel. (0228) 959.120
                            Telefax: (0228) 959.12.17

THAILAND – THAÏLANDE
Suksit Siam Co. Ltd.
113, 115 Fuang Nakhon Rd.
Opp. Wat Rajbopith
Bangkok 10200   Tel. (662) 225.9531/2
                            Telefax: (662) 222.5188

TURKEY – TURQUIE
Kültür Yayinlari Is-Türk Ltd. Sti.
Atatürk Bulvari No. 191/Kat 13
Kavaklidere/Ankara   Tel. 428.11.40 Ext. 2458
Dolmabahce Cad. No. 29
Besiktas/Istanbul   Tel. 260.71.88
                            Telex: 43482B

UNITED KINGDOM – ROYAUME-UNI
HMSO
Gen. enquiries   Tel. (071) 873 0011
Postal orders only:
P.O. Box 276, London SW8 5DT
Personal Callers HMSO Bookshop
49 High Holborn, London WC1V 6HB
                            Telefax: (071) 873 8200
Branches at: Belfast, Birmingham, Bristol,
Edinburgh, Manchester

UNITED STATES – ÉTATS-UNIS
OECD Publications and Information Center
2001 L Street N.W., Suite 650
Washington, D.C. 20036-4910 Tel. (202) 785.6323
                            Telefax: (202) 785.0350

VENEZUELA
Libreria del Este
Avda F. Miranda 52, Aptdo. 60337
Edificio Galipán
Caracas 106   Tel. 951.1705/951.2307/951.1297
                            Telegram: Libreste Caracas

Subscription to OECD periodicals may also be
placed through main subscription agencies.

Les abonnements aux publications périodiques de
l'OCDE peuvent être souscrits auprès des
principales agences d'abonnement.

Orders and inquiries from countries where Distribu-
tors have not yet been appointed should be sent to:
OECD Publications Service, 2 rue André-Pascal,
75775 Paris Cedex 16, France.

Les commandes provenant de pays où l'OCDE n'a
pas encore désigné de distributeur peuvent être
adressées à : OCDE, Service des Publications,
2, rue André-Pascal, 75775 Paris Cedex 16, France.

5-1995

OECD PUBLICATIONS, 2 rue André-Pascal, 75775 PARIS CEDEX 16
PRINTED IN FRANCE
(66 95 15 1) ISBN 92-64-14538-9 – No. 48094 1995